Skillmaths
2

Skillmaths 2

STAN A. ALTHANS, B.Sc.

formerly Senior Lecturer in Mathematics,
Newland Park College of Education, Buckinghamshire
now teacher of mathematics at St. Margaret's School,
Bushey, Herts

and

MARGERY DENT, B.Sc.

formerly Senior Lecturer in Mathematics,
Newland Park College of Education, Buckinghamshire

Macmillan Education
London and Basingstoke

First published 1983
Published by
MACMILLAN EDUCATION LIMITED
Houndmills Basingstoke Hampshire RG21 2XS
and London
Associated companies throughout the world

Typeset in Hong Kong by Asco Trade Typesetting Ltd.
Printed in Hong Kong

British Library Cataloguing in Publication Data
Althans, Stanley
Skillmaths 2
1. Mathematics – 1961–
I. Title II. Dent, Margery
510 QA39.2

ISBN 0-333-32223-1

Contents

Square and triangular numbers

A Copy and continue this pattern of squares, up to 8 × 8.

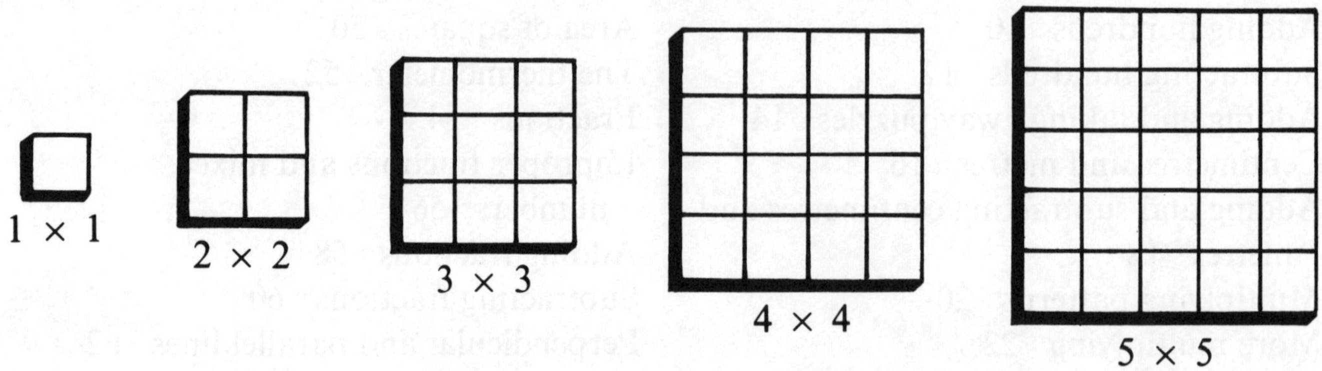

1 **4** **9** **16** **25**

These numbers are called SQUARE NUMBERS.

B SQUARE NUMBERS can also be built this way.
Continue the pattern below.

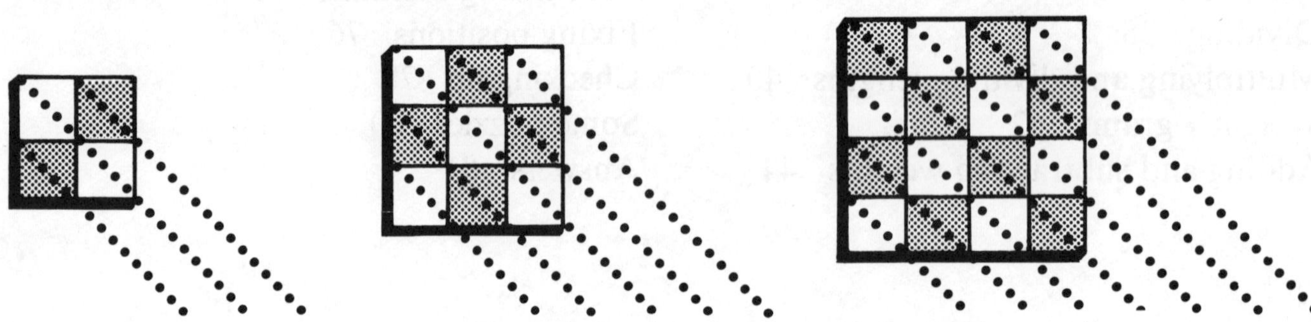

4 = 1 + 2 + 1 9 = 1 + 2 + 3 + 2 + 1 16 = 1 + 2 + 3 + 4 + 3 + 2 + 1

Complete these:

1. 1 + 2 + 3 + 4 + 5 + 4 + 3 + 2 + 1 = □ = □ × □
2. 1 + 2 + 3 + 4 + 5 + 6 + 5 + 4 + 3 + 2 + 1 = □ = □ × □
3. 1 + 2 + 3 + 4 + 5 + 6 + 7 + 6 + 5 + 4 + 3 + 2 + 1 = □ = □ × □

C List the set of SQUARE NUMBERS:
{1, 4, 9, 16, 25, ... 100}

D Draw squares on squared paper to find the SQUARE NUMBERS:
1. 11 × 11 **2.** 12 × 12 **3.** 13 × 13.

E These patterns make TRIANGULAR NUMBERS.

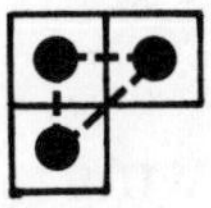

$1 + 2 = 3$

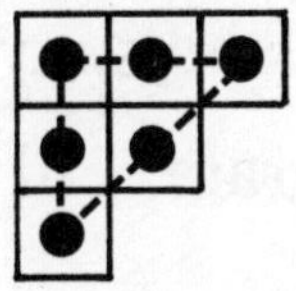

$1 + 2 + 3 = 6$

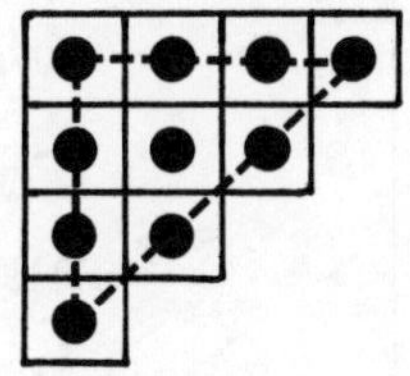

$1 + 2 + 3 + 4 = 10$

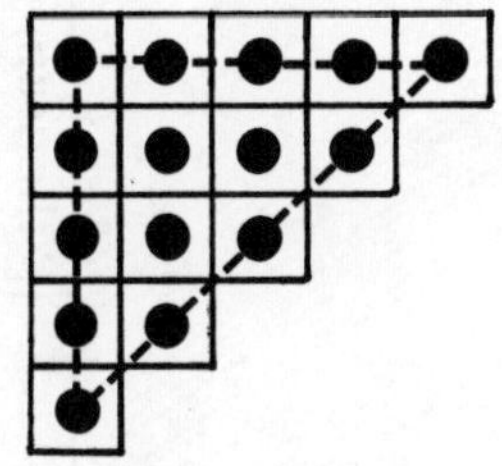

$1 + 2 + 3 + 4 + 5 = 15$

Draw the next two TRIANGULAR NUMBERS, and work out what they are.

Copy and complete these patterns:

1. $1 + 2 + 3 + 4 = \square$

2. $1 + 2 + 3 + 4 + 5 = \square$

3. $1 + 2 + 3 + 4 + 5 + 6 = \square$

4. $1 + 2 + 3 + 4 + 5 + 6 + 7 = \square$

5. $1 + 2 + 3 + 4 + 5 + 6 + 7 + \square = \square$

6. $1 + 2 + 3 + 4 + 5 + 6 + 7 + \square + \square = \square$

7. $1 + 2 + 3 + 4 + 5 + 6 + 7 + \square + \square + \square = \square$

F 1 is the first TRIANGULAR NUMBER.

List the set of the first ten TRIANGULAR NUMBERS.

{1, 3, 6, 10, 15, ...}

G SQUARE NUMBERS can be built from TRIANGULAR NUMBERS.

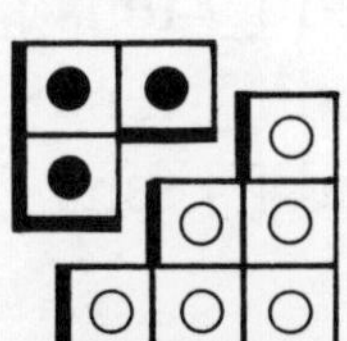

$3 + 6 = 9$
$3 \times 3 = 9$

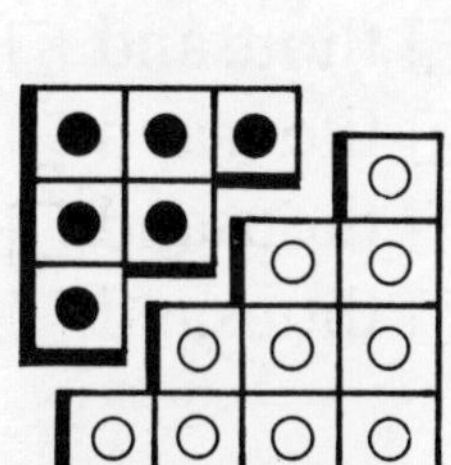

$6 + 10 = 16$
$4 \times 4 = 16$

$10 + 15 = 25$
$5 \times 5 = 25$

Make SQUARE NUMBERS from:

1. 15 and 21 **2.** 21 and 28

Thousands

A

thousands	hundreds	tens	ones
6	2	4	5
2	0	6	0

The two numbers in the diagram are:
six thousand two hundred and forty-five
two thousand and sixty

Copy the diagram. Fill in the following numbers:

1. five thousand two hundred and forty-six
2. four thousand and seventy
3. nine thousand and six
4. six thousand two hundred
5. seven thousand and eighty-eight

B

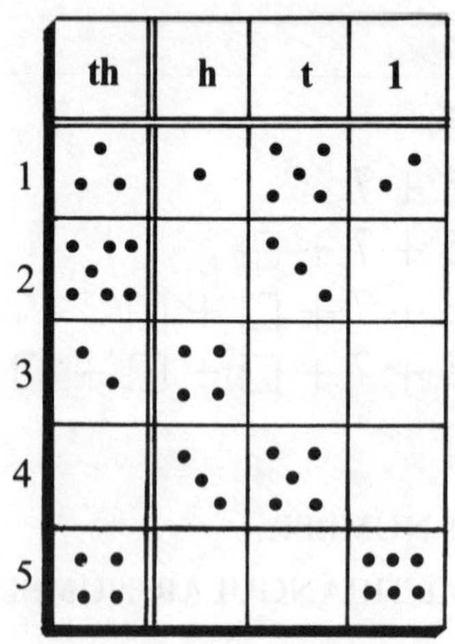

Write in figures the numbers shown on this counting board.

C

Copy the numbers and fill in the boxes:

1. 6520 → □ thousand □ hundred □ tens □ ones
2. 5044 → □ thousand □ hundred □ tens □ ones
3. 1256 → □ thousand □ hundred □ tens □ ones
4. 6206 → □ thousand □ hundred □ tens □ ones
5. 9471 → □ thousand □ hundred □ tens □ ones

D

Write the largest number you can make using:

1. 1, 7, 2, 3 **2.** 0, 0, 2, 5 **3.** 3, 0, 5, 3
4. 4, 7, 1, 9 **5.** 4, 8, 4, 6 **6.** 5, 2, 4, 3
7. 6, 1, 3, 4 **8.** 9, 2, 4, 7 **9.** 0, 4, 2, 7

E Write the largest number you can using four figures.

F Write the next:

1. five numbers after 997
2. five numbers after 1996
3. four numbers after 2498
4. eleven numbers after 2999
5. six numbers before 3003
6. six numbers before 1001
7. five numbers before 2704
8. twelve numbers before 3010.

G Write the number that is:

1. 2 more than 999
2. 4 more than 2996
3. 3 more than 2999
4. 11 more than 5999
5. 2 less than 5000
6. 3 less than 5001
7. 12 less than 5010
8. 30 less than 2020.

H Put these rivers in order according to their lengths, put the longest one first.

R. Nile	6324 km	R. Mississippi	6051 km	R. Danube	2850 km
R. Amazon	6518 km	R. Rhine	1320 km	R. Volga	3668 km

I The points scored in the decathlon (10 athletics events) were:

B. Rush	3029	J. Bull	4001
A. Lamb	3902	C. Side	3999
N. Parker	4100	T. Mann	3072

Put the names in the order they came –
1st, 2nd, 3rd, 4th, 5th, 6th.

J Copy and complete the number line:

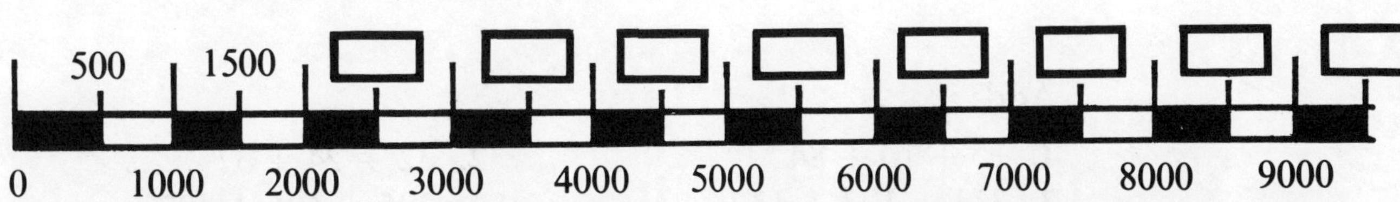

1. Is 750 nearer to 0 or to 1000?
2. Is 1670 nearer to 1000 or to 2000?
3. Is 2639 nearer to 2000 or to 3000?
4. Is 5275 nearer to 5000 or to 6000?
5. Is 7099 nearer to 7000 or to 8000?
6. Is 9603 nearer to 9000 or to 10 000?

Adding hundreds

A *Examples:*

1.	2.	3.	4.
72	368	453	326
+ 39	+ 226	+ 834	+ 284
111	**594**	**1287**	**610**

Find these sums.

1. 51 + 78	**2.** 42 + 78	**3.** 95 + 59	**4.** 88 + 67	**5.** 49 + 71
6. 177 + 82	**7.** 259 + 63	**8.** 27 + 179	**9.** 240 + 67	**10.** 323 + 87
11. 174 + 26	**12.** 356 + 47	**13.** 698 + 183	**14.** 372 + 458	**15.** 527 + 279
16. 497 + 315	**17.** 599 + 149	**18.** 655 + 275	**19.** 79 + 824	**20.** 628 + 233
21. 182 + 430	**22.** 694 + 205	**23.** 350 + 750	**24.** 909 + 100	**25.** 280 + 830
26. 581 + 476	**27.** 949 + 651	**28.** 555 + 445	**29.** 999 + 999	**30.** 678 + 876

B Add:

1. 15 + 10 + 84
2. 32 + 24 + 37
3. 311 + 123 + 132
4. 205 + 406 + 402
5. 128 + 409 + 130
6. 345 + 156 + 27
7. 400 + 201 + 17
8. 824 + 6 + 42
9. 33 + 9 + 602
10. 20 + 8 + 450

C

A salesman kept this record of his car journeys, in kilometres (km).

WEEK	MONDAY	TUESDAY	WEDNESDAY	THURSDAY	FRIDAY	TOTALS
1	74	9	37	127	10	
2	28	12	42	109	8	
3	32	8	35	113	8	
TOTALS						

1. Work out the total distances travelled each week.
2. Work out the total distances travelled on Mondays, Tuesdays, Wednesdays, Thursdays, Fridays.
3. Work out the total distance travelled overall.
4. Which day did he travel farthest?
5. Which week did he do most travelling?

D

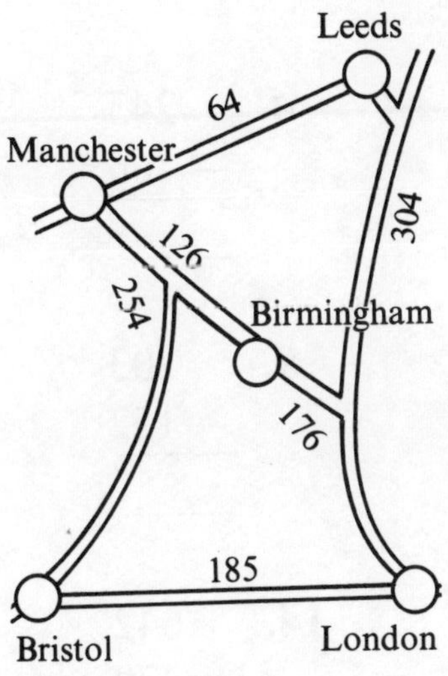

The distances on the map are in kilometres (km).
Work out the lengths of these journeys:

1. London to Birmingham to Manchester
2. Bristol to Manchester to Leeds
3. Bristol to London to Birmingham
4. Manchester to Leeds to London.
5. Birmingham to Manchester to Leeds
6. Birmingham to London to Bristol to Manchester

E

These are the numbers in the classes at Highwood School.

1st Year		2nd Year		3rd Year		4th Year	
Class 1 28	Class 2 29	Class 3 28	Class 4 31	Class 5 29	Class 6 33	Class 7 29	Class 8 29

1. How many children are there in each year?
2. How many children are there at Highwood School?

Subtracting hundreds

A *Example:*

$$\begin{array}{r} 432 \\ -\,257 \\ \hline \\ \hline \end{array} \rightarrow \begin{array}{r} (400+30+2) \\ -\,(200+50+7) \\ \hline \\ \hline \end{array} \rightarrow \begin{array}{r} (400+20+12) \\ -\,(200+50+\ \ 7) \\ \hline 5 \\ \hline \end{array} \rightarrow \begin{array}{r} (300+120) \\ -\,(200+\ \ 50) \\ \hline \mathbf{100+\ \ 70+5\rightarrow 175} \\ \hline \end{array}$$

OR

$$\begin{array}{r} (400+30+2) \\ -\,(200+50+7) \\ \hline \\ \hline \end{array} \rightarrow \begin{array}{r} (400+30+12) \\ -\,(200+60+\ \ 7) \\ \hline 5 \\ \hline \end{array}$$

I have added 10 to the top line and 10 to the bottom line

$$\begin{array}{r} (400+130) \\ -\,(300+\ \ 60) \\ \hline \mathbf{100+\ \ 70+5} \end{array}$$

I have added 100 to the top and bottom lines

175

Work out:

1. $\begin{array}{r} 452 \\ -\ \ 47 \\ \hline \\ \hline \end{array}$ **2.** $\begin{array}{r} 629 \\ -\ \ 46 \\ \hline \\ \hline \end{array}$ **3.** $\begin{array}{r} 127 \\ -\ \ 89 \\ \hline \\ \hline \end{array}$ **4.** $\begin{array}{r} 243 \\ -\ \ 66 \\ \hline \\ \hline \end{array}$ **5.** $\begin{array}{r} 352 \\ -\,168 \\ \hline \\ \hline \end{array}$

6. $\begin{array}{r} 273 \\ -\,184 \\ \hline \\ \hline \end{array}$ **7.** $\begin{array}{r} 524 \\ -\,456 \\ \hline \\ \hline \end{array}$ **8.** $\begin{array}{r} 416 \\ -\,109 \\ \hline \\ \hline \end{array}$ **9.** $\begin{array}{r} 203 \\ -\,152 \\ \hline \\ \hline \end{array}$ **10.** $\begin{array}{r} 410 \\ -\,183 \\ \hline \\ \hline \end{array}$

11. $\begin{array}{r} 630 \\ -\,149 \\ \hline \\ \hline \end{array}$ **12.** $\begin{array}{r} 506 \\ -\,234 \\ \hline \\ \hline \end{array}$ **13.** $\begin{array}{r} 974 \\ -\,309 \\ \hline \\ \hline \end{array}$ **14.** $\begin{array}{r} 642 \\ -\,470 \\ \hline \\ \hline \end{array}$ **15.** $\begin{array}{r} 704 \\ -\ \ 73 \\ \hline \\ \hline \end{array}$

16. $\begin{array}{r} 605 \\ -\ \ 89 \\ \hline \\ \hline \end{array}$ **17.** $\begin{array}{r} 405 \\ -\ \ 96 \\ \hline \\ \hline \end{array}$ **18.** $\begin{array}{r} 702 \\ -\ \ 97 \\ \hline \\ \hline \end{array}$ **19.** $\begin{array}{r} 500 \\ -\,174 \\ \hline \\ \hline \end{array}$ **20.** $\begin{array}{r} 900 \\ -\,399 \\ \hline \\ \hline \end{array}$

21. 369 − 213 **22.** 427 − 173 **23.** 589 − 97

24. 423 − 67 **25.** 503 − 174 **26.** 918 − 59

B Find the difference between:

1. 629 and 34 **2.** 427 and 79 **3.** 54 and 176
4. 48 and 129 **5.** 625 and 125 **6.** 476 and 724
7. 508 and 576 **8.** 329 and 606 **9.** 252 and 500.

C What must I add

1. to 42 to make 168? **2.** to 74 to make 144?
3. to 192 to make 365? **4.** to 108 to make 180?
5. to 265 to make 360? **6.** to 492 to make 999?
7. to 196 to make 366? **8.** to 504 to make 720?

D Fill in the missing numbers:

1. $\begin{array}{r} 29 \\ +\square\square \\ \hline 112 \\ \hline \end{array}$ **2.** $\begin{array}{r} 420 \\ +\square\square\square \\ \hline 760 \\ \hline \end{array}$ **3.** $\begin{array}{r} 729 \\ +\square\square\square \\ \hline 972 \\ \hline \end{array}$ **4.** $\begin{array}{r} 625 \\ +\square\square\square \\ \hline 814 \\ \hline \end{array}$

5. $\begin{array}{r} 74 \\ +\square\square\square \\ \hline 472 \\ \hline \end{array}$ **6.** $\begin{array}{r} \square\square \\ +142 \\ \hline 194 \\ \hline \end{array}$ **7.** $\begin{array}{r} \square\square\square \\ +324 \\ \hline 472 \\ \hline \end{array}$ **8.** $\begin{array}{r} \square\square\square \\ +509 \\ \hline 720 \\ \hline \end{array}$

9. $\begin{array}{r} \square\square \\ +192 \\ \hline 241 \\ \hline \end{array}$ **10.** $\begin{array}{r} \square\square\square \\ +63 \\ \hline 441 \\ \hline \end{array}$

E Work out this puzzle:

	1.	**2.**	**3.**	**4.**
	624	924	321	623
written backwards	− 426	− 429	− 123	− 326
	198	495	198	□□7
written backwards	+ 891	+ 594	+ □□□	+ 7□□

	5.	**6.**	**7.**	**8.**
	591	732	821	992
written backwards	− □□5	− 2□□	− □□□	− □□□
	□□□	□□□	□□□	□□□
written backwards	+ □□□	+ □□□	+ □□□	+ □□□

Make up some yourself.

Adding and taking away puzzles

A These are VANISHING TRIANGLES. Complete them.

1.

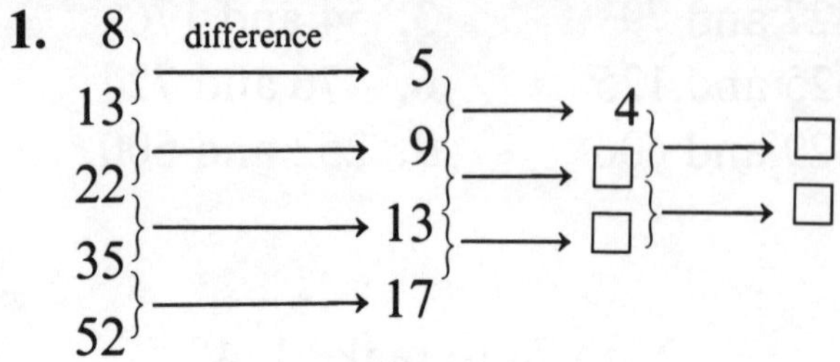

2.

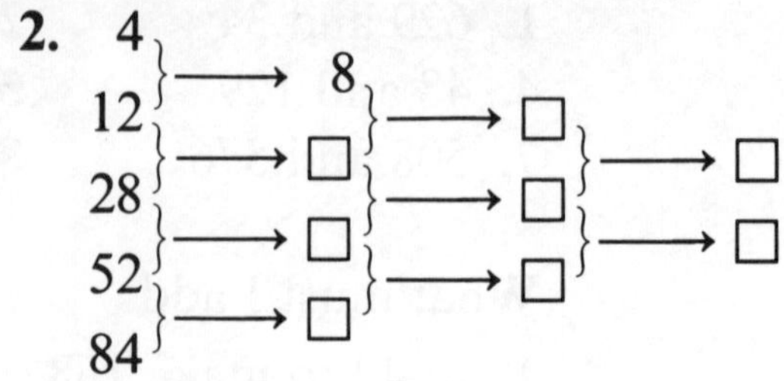

B Copy and complete the cross-number puzzle.

Clues across
- **1** 32 − 9
- **4** 28 + 17
- **6** 50 − 24
- **8** 38 + 24
- **10** 81 − 27
- **12** 15 + 17

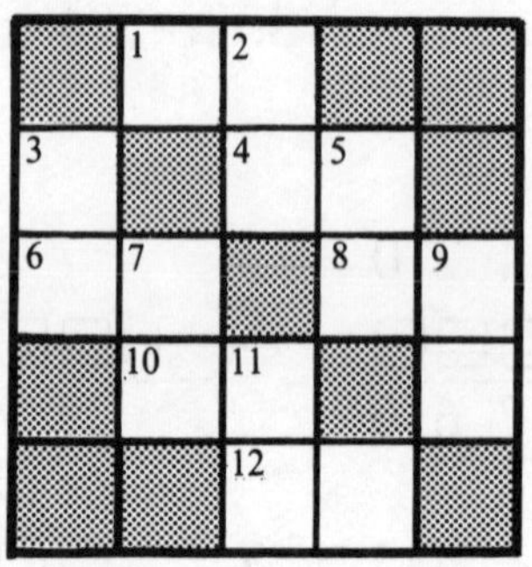

Clues down
- **2** 63 − 29
- **3** 36 + 36
- **5** 80 − 24
- **7** 38 + 27
- **9** 54 − 27
- **11** 25 + 18

C

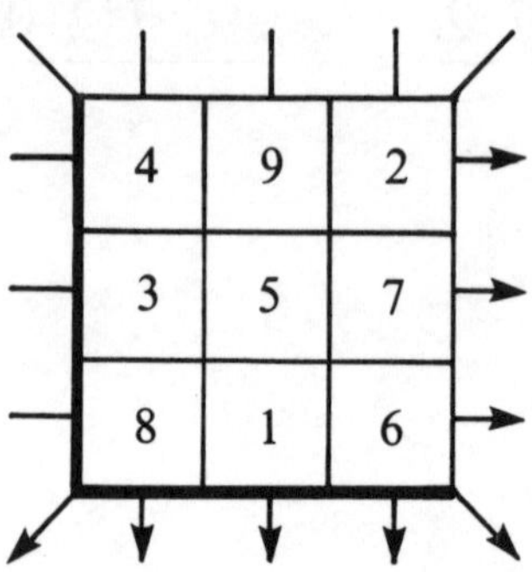

4 + 9 + 2 = □

This is a MAGIC SQUARE.
Add the numbers in all directions.
Find the MAGIC NUMBER.

Complete the MAGIC SQUARES below.

1.

5	10	3
9	2	7

Magic Number **18**

2.

10		6
	12	
		14

Magic Number **36**

3.

24		
12		6

Magic Number **45**

4.

		7
		17
19		

Magic Number **39**

D

The schools in Turlandshire have the following numbers of pupils.
Copy and complete the table.

SCHOOL	Wigton	Cedars	Roxburgh	Belmont	St. Peter's	Culver
boys girls	394 427	292 318	426	 349	245	592 624
TOTAL			860	688	500	

E

In a Test Match, the scores were:

England 1st innings 278 — 2nd innings 358
Australia 1st innings 294 — 2nd innings 349

1. Work out England's total score.
2. Work out Australia's total score.
3. Who led on the 1st innings, and by how many runs?
4. Which side scored more in the 2nd innings? How many more runs?
5. Which team won, and by how many runs?
6. How many runs were scored in the match?

F

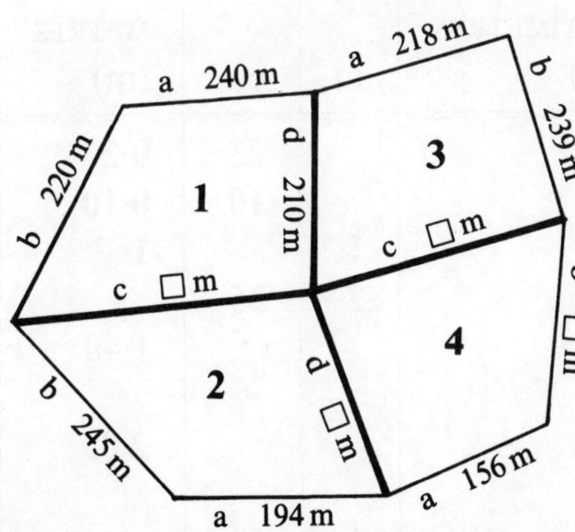

The perimeter (distance all round) of each of Farmer Brown's fields is 990 m (metres). Work out the lengths of the sides of each field to complete the table.

FIELD	1	2	3	4
side a side b side c side d	240 m 220 m 210 m	194 m 245 m	218 m 239 m 210 m	156 m
PERIMETER	990 m	990 m	990 m	990 m

Centimetres and metres

A 100 CENTIMETRES (cm) = 1 METRE (m).
1 centimetre is $\frac{1}{100}$ (one hundredth) of a metre.

We write:
1 cm → **0·01 m**
10 cm → **0·10 m**
25 cm → **0·25 m**

Copy and complete the table:

cm	1	2	3	4	5	6	7	8	9
m	0·01	0·02							

cm	10	11	12	13	14	15	16	17	18	19
m	0·10	0·11								

cm	20	30	40	50	60	70	80	90	100
m	0·20	0·30							

cm	110	120	130	140	150	160	170	180	190	200
m	1·10	1·20								

B Copy and complete this table:

CENTIMETRES (cm)	m	cm	METRES (m)
25		**25**	**0·25**
10		**10**	**0·10**
125	**1**		**1·**
	2	35	
			1·40
264			·
			2·05
309			·
	4	50	·
7			·
400			·
	2	6	·

C

Write in CENTIMETRES (cm).

1. 0·75 m	**2.** 0·62 m	**3.** 0·40 m	**4.** 0·09 m
5. 0·07 m	**6.** 0·05 m	**7.** 0·08 m	**8.** 0·02 m
9. 1·00 m	**10.** 1·07 m	**11.** 1·70 m	**12.** 1·08 m
13. 1·80 m	**14.** 1·06 m	**15.** 1·60 m	**16.** 2·75 m
17. 4·72 m	**18.** 3·05 m	**19.** 6·09 m	**20.** 3·89 m

D

Write as METRES (m). Number 1 is done for you.

1. 125 cm → **1·25 m**

2. 246 cm	**3.** 127 cm	**4.** 107 cm	**5.** 205 cm
6. 150 cm	**7.** 186 cm	**8.** 110 cm	**9.** 101 cm
10. 53 cm	**11.** 74 cm	**12.** 60 cm	**13.** 6 cm
14. 40 cm	**15.** 4 cm	**16.** 9 cm	**17.** 90 cm

E

Write as METRES AND CENTIMETRES (m cm). Number 1 is done for you.

1. 1·63 m → **1m 63 cm**

2. 1·48 m	**3.** 1·26 m	**4.** 2·80 m	**5.** 2·08 m
6. 2·00 m	**7.** 1·60 m	**8.** 1·06 m	**9.** 3·70 m
10. 3·07 m	**11.** 0·90 m	**12.** 0·20 m	**13.** 0·30 m
14. 0·03 m	**15.** 0·05 m	**16.** 0·50 m	**17.** 1·50 m

F

Would you use METRES (m) or CENTIMETRES (cm) to measure these?

1. the length of your classroom
2. the length of your exercise book
3. the height of your classroom
4. the length of the playground
5. the spread of your fingers
6. the height of a dog
7. the width of a TV screen.

Adding and subtracting centimetres and metres

A Work out:

1.	**2.**	**3.**	**4.**
m cm	m cm	m cm	m cm
1 54	2 64	7 28	5 29
+ 3 72	+ 1 73	+ 1 62	+ 71
5 26			
1			

5.	**6.**	**7.**	**8.**
m cm	m cm	m cm	m cm
6 49	57	2 6	3 76
+ 82	+ 1 56	+ 1 95	+ 2 67

9.	**10.**	**11.**	**12.**
m	m	m	m
1·48	3·57	6·05	8·67
+ 2·61	+ 1·93	+ 0·95	+ 5·40
4·09			

13.	**14.**	**15.**	**16.**
m	m	m	m
8·03	3·05	1·09	4·75
+ 1·77	+ 0·79	+ 2·01	+ 5·25

B

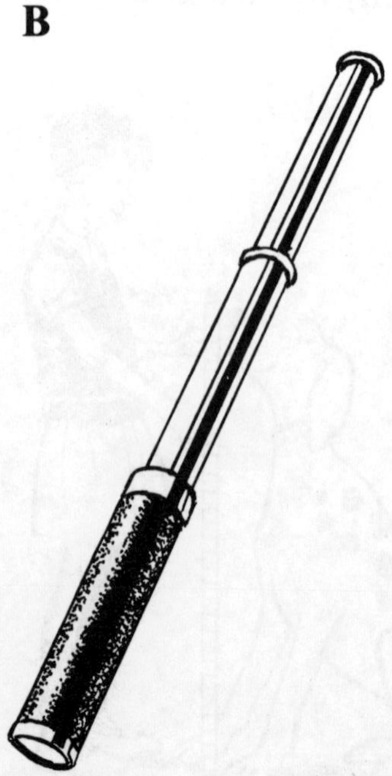

1. A telescope opens out into 3 lengths of 45 cm, 40 cm and 35 cm.
How long is it when fully opened
(a) in centimetres?
(b) in metres and centimetres?

2. Ian made 3 giant strides, 1·46 m, 1·58 m, 1·23 m.
(a) How far had he jumped after 2 strides?
(b) How far did he go altogether?

3. Paul jumped 4·26 m. The school record is 84 cm longer.
How far must he jump to equal the record,
(a) in metres?
(b) in metres and centimetres?

C Find the difference:

1.	2.	3.	4.
m cm	m cm	m cm	m cm
3 28	6 59	5 24	2 94
− 1 82	− 2 68	− 3 96	− 1 76

5.	6.	7.	8.
m cm	m cm	m cm	m cm
2 48	5 08	1 18	6 42
− 1 67	− 99	− 74	− 1 6

9.	10.	11.	12.
m	m	m	m
5·68	3·84	1·26	4·08
− 2·72	− 0·76	− 0·79	− 1·24

13.	14.	15.	16.
m	m	m	m
8·24	6·36	1·76	4·28
− 7·49	− 5·48	− 0·29	− 4·24

D

1. Peter is 1·48 m tall. His mother is 12 cm taller. His father is 1·76 m tall.
   (a) How tall is Peter's mother?
   (b) How much shorter is Peter than his father?
   (c) What is the difference between his mother's and his father's heights?

2. A garage is 2·55 m wide.
   A car parked inside has 45 cm of space one side, and 53 cm the other side.
   (a) How much space is there on both sides of the car altogether?
   (b) How wide is the car?
   (c) If the car were in the middle, how much space would there be on either side?

Multiplying patterns

A Complete these sequences of 10's.

1.
$2 \times 4 + 2 = 10$
$4 \times 4 + 4 = \square$
$6 \times 4 + 6 = \square$
$8 \times 4 + \square = \square$
$\square \times 4 + \square = \square$
$\square \times 4 + \square = \square$
$\square \times 4 + \square = \square$

2.
$1 \times 6 + 4 = 10$
$3 \times 6 + 2 = \square$
$5 \times 6 + 0 = \square$
$\square \times 6 - 2 = \square$
$\square \times 6 - \square = \square$
$\square \times 6 - \square = \square$
$\square \times 6 - \square = \square$

B Continue these multiplying sequences.

1.
$1 \times 8 + 1 = 9$
$12 \times 8 + 2 = 9\square$
$123 \times 8 + 3 = 9\square\square$
$1234 \times 8 + 4 = 9\square\square\square$
and so on

2.
$1 \times 7 + 1 = 8$
$12 \times 7 + 2 = 8\square$
$123 \times 7 + 3 = 8\square\square$
$1234 \times 7 + 4 = 8\square\square\square$
and so on

C 37 is a peculiar number.

1. $37 \times 3 = \square$ **2.** $37 \times 6 = \square$ **3.** $37 \times 9 = \square$

Can you continue the pattern?

D Try these:

1.
$2 \times 3 - 1 \times 4 = \square$
$3 \times 4 - 2 \times 5 = \square$
$4 \times 5 - 3 \times 6 = \square$
$5 \times \square - 4 \times \square = 2$
$\square \times 7 - \square \times 8 = 2$
$\square \times \square - 6 \times 9 = 2$

2.
$3 \times 4 - 1 \times 5 = \square$
$4 \times 5 - 2 \times 6 = \square$
$5 \times 6 - 3 \times 7 = \square$
$6 \times 7 - 4 \times 8 = \square$
$7 \times 8 - 5 \times 9 = \square$
$8 \times 9 - 6 \times 10 = \square$

E

1. Build up the 7-times table.

$$7 + 7 = 2 \times 7 = 14$$
$$7 + 7 + 7 = 3 \times 7 = \square$$
$$7 + 7 + 7 + 7 = 4 \times 7 = \square$$
$$7 + 7 + 7 + 7 + 7 = 5 \times 7 = \square$$
$$7 + 7 + 7 + 7 + 7 + 7 = 6 \times 7 = \square$$

Continue up to 10 × 7.

2. Build up the 4, 5, 6, 8, 9 times tables in the same way.

F Copy and complete the multiplication square.

10										
9										
8	→	→	→	→	40					
7					↑					
6					↑		42			
5	→	→	15		↑					
4			↑		↑					
3			↑		↑					
2	→	4	↑		↑					
1		↑	↑		↑					
×	1	2	3	4	5	6	7	8	9	10

Answer the following with or without using the square.

1. 3 × 5	**2.** 2 × 2	**3.** 6 × 7
4. 5 × 8	**5.** 4 × 3	**6.** 7 × 4
7. 4 × 7	**8.** 9 × 6	**9.** 6 × 9
10. 5 × 5	**11.** 7 × 3	**12.** 6 × 4
13. 7 × 8	**14.** 8 × 7	**15.** 8 × 8
16. 2 × 9	**17.** 7 × 2	**18.** 9 × 7
19. 7 × 9	**20.** 10 × 10	**21.** 4 × 8
22. 8 × 9	**23.** 9 × 8	**24.** 6 × 8
25. 2 × 8	**26.** 8 × 3	**27.** 9 × 10

G

Multiply the following:

1. 3 × 5; 7 × 3; 5 × 9; 9 × 7
2. 2 × 4; 6 × 8; 4 × 8; 8 × 2
3. 5 × 4; 3 × 6; 5 × 8; 7 × 6

Now complete the table:

×	ODD	EVEN
ODD		
EVEN		

More multiplying

A $10 \times 10 \times 10 \rightarrow (10 \times 10) \times 10 \rightarrow 100 \times 10 = 1000.$

Work out:

1. $(2 \times 2) \times 2 = \square$	**2.** $3 \times 3 \times 3 = \square$	**3.** $4 \times 4 \times 4 = \square$
4. $2 \times 2 \times 2 \times 2 = \square$	**5.** $3 \times 3 \times 3 \times 3 = \square$	**6.** $5 \times 5 \times 5 = \square$
7. $6 \times 6 \times 6 = \square$	**8.** $5 \times 5 \times 5 \times 5 = \square$	**9.** $7 \times 7 \times 7 = \square$
10. $8 \times 8 \times 8 = \square$	**11.** $9 \times 9 \times 9 = \square$	**12.** $4 \times 4 \times 4 \times 4 = \square$

B Work out:

1. $(2 \times 3) \times 4 = \square$	**2.** $3 \times 4 \times 5 = \square$	**3.** $4 \times 5 \times 6 = \square$
4. $5 \times 6 \times 7 = \square$	**5.** $6 \times 7 \times 8 = \square$	**6.** $7 \times 8 \times 9 = \square$
7. $8 \times 9 \times 10 = \square$	**8.** $0 \times 1 \times 2 = \square$	**9.** $8 \times 9 \times 0 = \square$
10. $1 \times 0 \times 3 = \square$	**11.** $5 \times 7 \times 9 = \square$	**12.** $2 \times 4 \times 6 = \square$

> **REMEMBER:** if a *number* is MULTIPLIED BY NOUGHT the *answer* is NOUGHT

C Copy and complete the arrow graphs.
The arrows tell you what to do.

1.

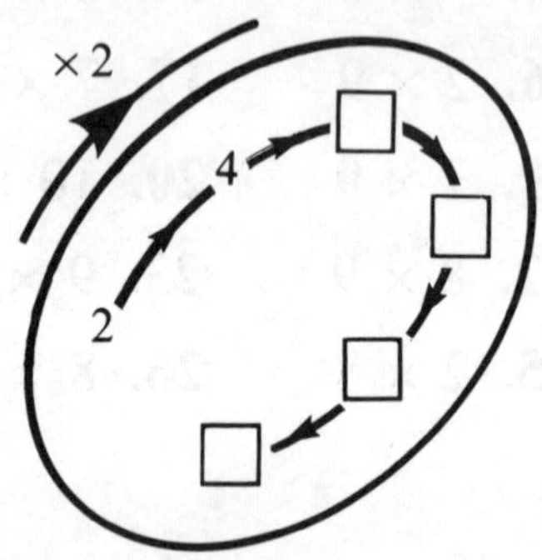

2.

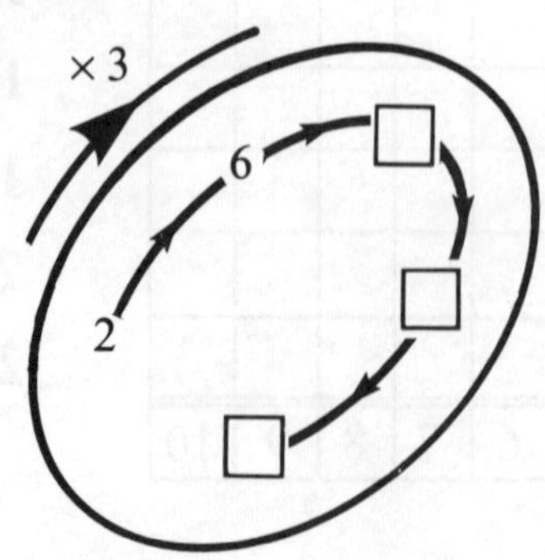

3.

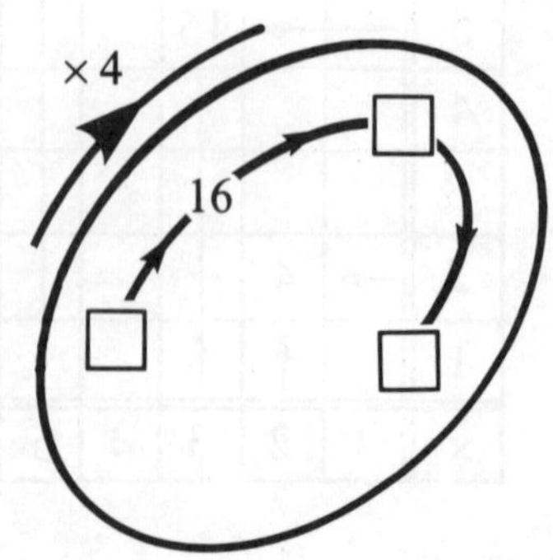

4.

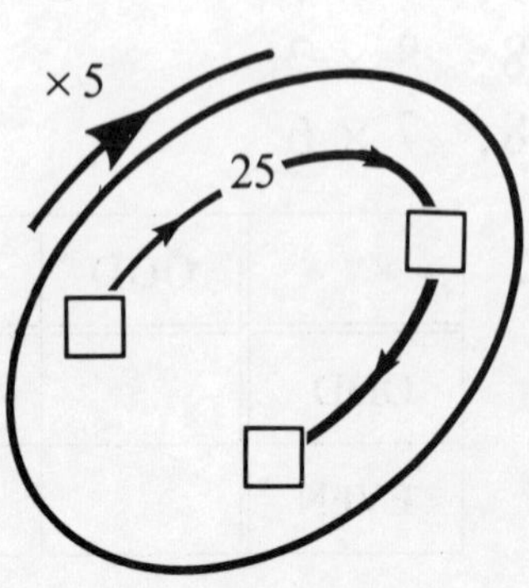

5.

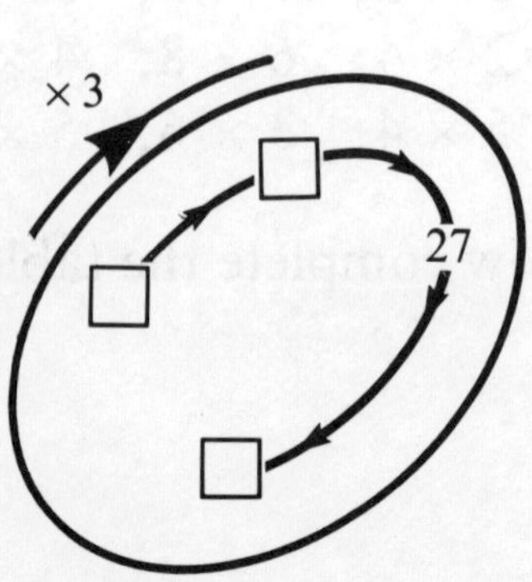

6.

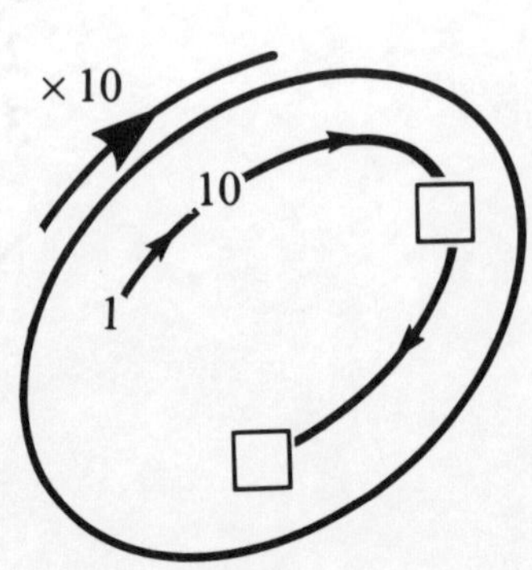

D *Examples:* **1.** 126 × 4 **2.** 207 × 3

1. $\begin{array}{r} 126 \\ \times\ \ 4 \\ \hline \mathbf{504} \end{array}$ **2.** $\begin{array}{r} 207 \\ \times\ \ 3 \\ \hline \mathbf{621} \end{array}$

Work out:

1. 27 × 4 **2.** 104 × 6 **3.** 108 × 7 **4.** 147 × 3 **5.** 186 × 4

6. 93 × 5 **7.** 287 × 2 **8.** 177 × 5 **9.** 176 × 4 **10.** 251 × 3

11. 158 × 0 **12.** 447 × 2 **13.** 529 × 2 **14.** 444 × 4 **15.** 425 × 7

E Multiplying by 10.

1. 3 × 10 = □ **2.** 7 × 10 = □ **3.** 4 × 10 = □ **4.** 12 × 10 = □
5. 11 × 10 = □ **6.** 17 × 10 = □ **7.** 14 × 10 = □ **8.** 13 × 10 = □
9. 21 × 10 = □ **10.** 20 × 10 = □ **11.** 25 × 10 = □ **12.** 30 × 10 = □
13. 38 × 10 = □ **14.** 33 × 10 = □ **15.** 46 × 10 = □ **16.** 50 × 10 = □
17. 64 × 10 = □ **18.** 99 × 10 = □ **19.** 90 × 10 = □ **20.** 100 × 10 = □

Have you discovered a quick way to multiply by 10?

F Complete the mappings.

1.

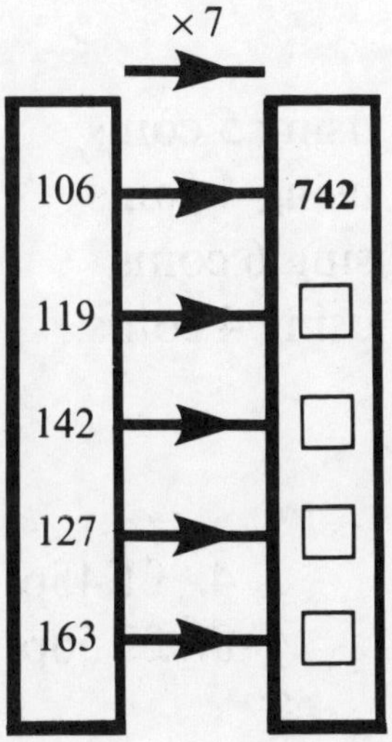

2.

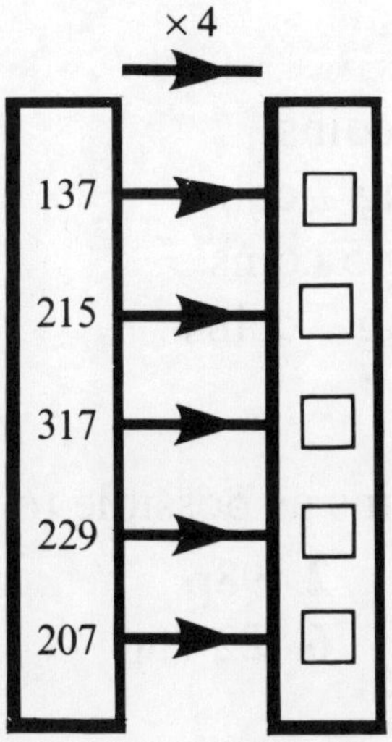

3.

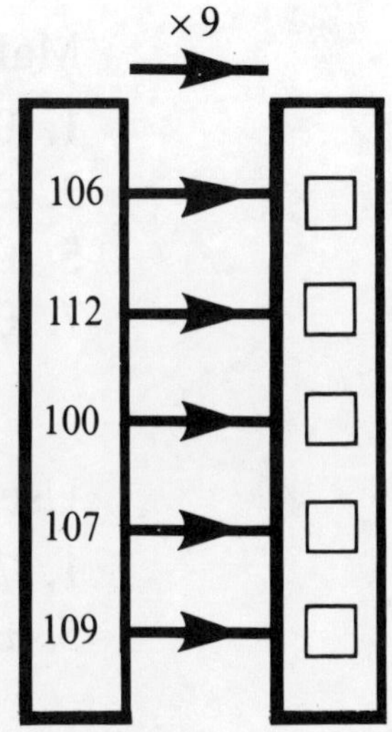

Money – pounds and pence

A

2 fifty pence = £1 (one pound)
100 pence = £1

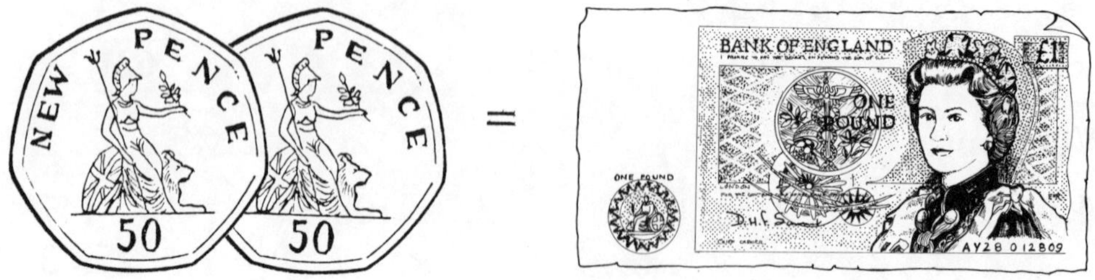

1. How many of these make £1?
(a) 20p (b) 10p (c) 5p (d) 2p (e) 1p

2. Which single coin is:
(a) £$\frac{1}{2}$ (b) £$\frac{1}{10}$ (c) £$\frac{1}{20}$ (d) £$\frac{1}{50}$ (e) £$\frac{1}{100}$?

B

Work out the value of these coins:

1. 2 fifty pence and 1 ten pence
2. 1 fifty pence, 2 twenty pence and 2 five pence
3. 3 fifty pence
4. 3 fifty pence and 2 ten pence
5. 5 ten pence, 10 five pence and 3 two pence
6. 1 fifty pence, 1 twenty pence and 7 ten pence
7. 4 fifty pence, 10 five pence and 5 two pence
8. 5 fifty pence and 5 five pence.

C

Make up:

1. £1 using 6 coins
2. £1 17p using 5 coins
3. £2 13p using 7 coins
4. £1 55p using 4 coins
5. £1 8p using 5 coins
6. £2 7p using 6 coins
7. £2 65p using 7 coins
8. £1 11p using 4 coins.

D

Use as few coins as possible to make:

1. 72p **2.** 98p **3.** £1 5p **4.** £1 45p
5. £2 15p **6.** £2 75p **7.** £3 6p **8.** £3 56p.

E Money on cheques must be written in a special way.

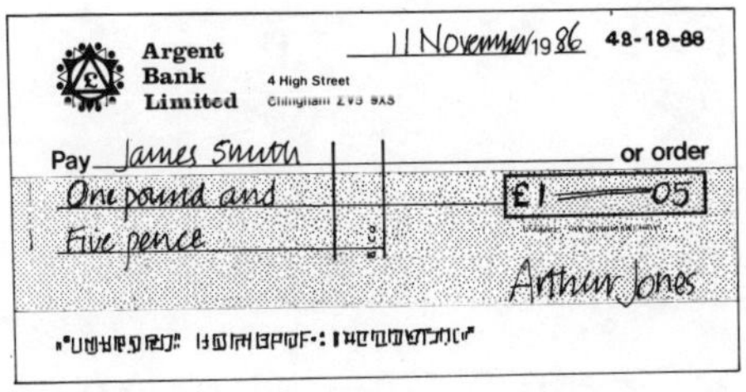

One pound and five pence (£1 5p) is written **£1–05**

Copy these cheques and complete them:

1.

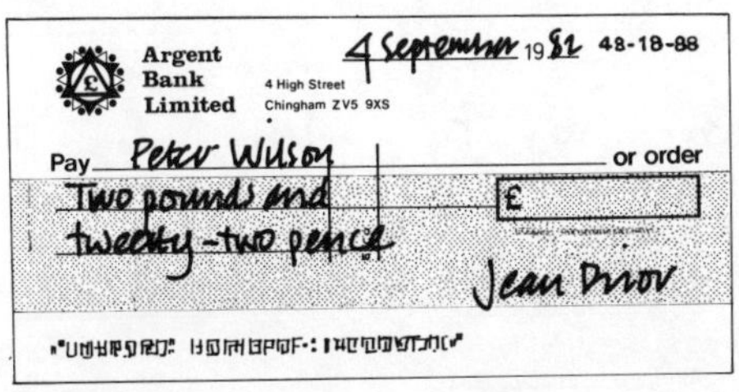

2.

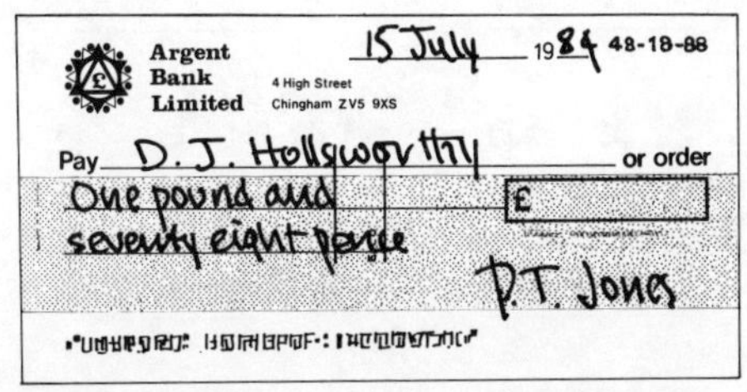

3.

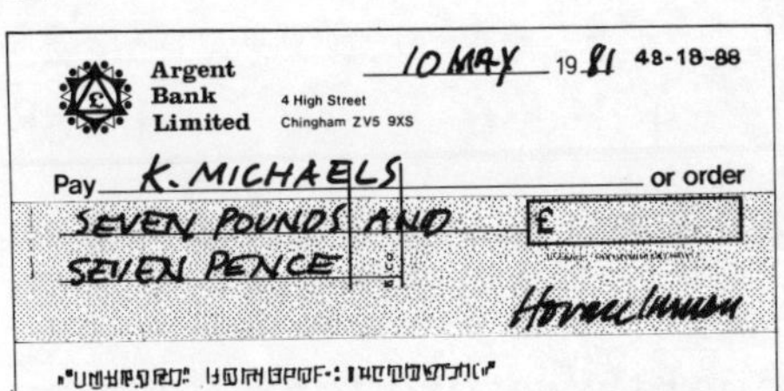

4.

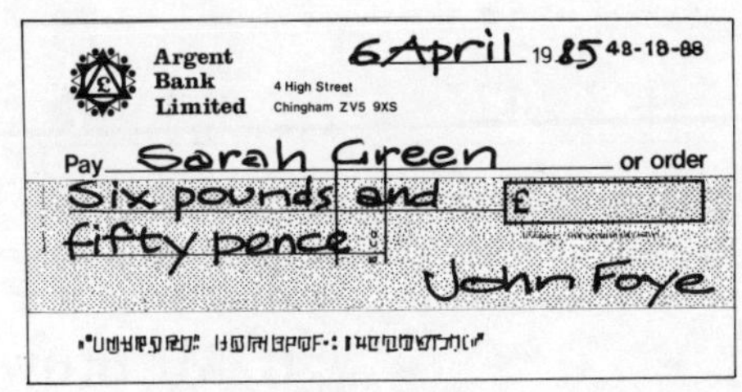

F £1 37p can be written as **£1·37**.
47p is written as **£0·47**. £1 5p is written as **£1·05**.
Copy and complete the table:

pence	£	p	£
125	**1**	**25**	**1·25**
105			**1·05**
	1	72	
			2·06
417			
	3	5	
74			
	2	9	

Adding money

A Work out:

1. p	**2.** p	**3.** p	**4.** p	**5.** p
74	56	32	65	35
+ 16	+ 18	+ 49	+ 47	+ 55

6. £ p	**7.** £ p	**8.** £ p	**9.** £ p	**10.** £ p
54	48	75	89	97
+ 63	+ 63	+ 25	+ 98	+ 9
1 17				

11. £ p	**12.** £ p	**13.** £ p	**14.** £ p	**15.** £ p
1 47	2 53	58	2 64	4 2
+ 28	+ 65	+ 1 70	+ 1 37	+ 98

16. £ p	**17.** £ p	**18.** £ p	**19.** £ p	**20.** £ p
1 25	75	5 64	3 25	6 47
+ 3 42	+ 1 45	+ 1 36	+ 1 88	+ 3 53

B

1. Mr Brown bought a record for £2 99p and a tape for £1 40p. At another shop he bought a calculator for £14 99p.
(a) How much did Mr Brown spend at the first shop?
(b) How much did he spend altogether?

2. Mr Brown then took his wife to the theatre. They went for a meal first; that cost him £9 75p. The theatre tickets were £4 25p each.
(a) How much did Mr Brown pay for the tickets?
(b) How much did he spend altogether?

3. An advertisement in a newspaper cost £1 45p for a week. How much would it cost for (a) 2 weeks? (b) 3 weeks?

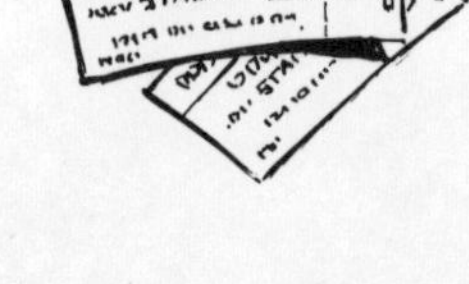

C

A bank cashier read his cheques, £1·95 and £2·07.

To add them up he wrote:

```
   £
  1·95
+ 2·07
  4·02
  1 1
```

Add these values:

1. £ 1·85 + 1·13 = .
2. £ 0·97 + 2·06 = .
3. £ 4·72 + 1·09 = .
4. £ 0·87 + 4·48 = .
5. £ 0·84 + 0·77 = .
6. £ 1·05 + 2·77 = .
7. £ 6·78 + 3·22 = .
8. £ 7·55 + 0·95 = .
9. £ 4·24 + 3·86 = .
10. £ 0·68 + 3·95 = .
11. £ 0·79 + 0·43 = .
12. £ 5·82 + 2·18 = .

13. £5·42 + £3·59
14. £6·05 + £0·95
15. £1·99 + £0·09
16. £5·76 + £4·24

D

1. A garage mechanic charged £5·77 for parts and £12·80 for labour. Work out the total bill.
2. A shop-keeper bought some goods. Complete the table he made.

COST PRICE	PROFIT	SELLING PRICE
£4·60	£0·80	**£5·40**
£1·25	£0·42	
£3·87	£1·13	
£9·05	£3·99	

Subtracting money

A Work out:

1.	**2.**	**3.**	**4.**	**5.**
p	p	p	p	p
25	23	36	42	53
− 13	− 18	− 27	− 29	− 19

6.	**7.**	**8.**	**9.**	**10.**
£ p	£ p	£ p	£ p	£ p
1 25	1 32	2 84	3 0	4 28
− 17	− 26	− 1 29	− 1 99	− 1 16

11.	**12.**	**13.**	**14.**	**15.**
£ p	£ p	£ p	£ p	£ p
3 44	3 52	4 28	8 0	10 0
− 1 28	− 1 57	− 2 99	− 3 85	− 7 15

16.	**17.**	**18.**	**19.**	**20.**
£ p	£ p	£ p	£ p	£ p
6 5	5 9	10 6	6 4	12 0
− 1 75	− 2 48	− 7 28	− 5 68	− 4 62

B

1. Tracey bought a chocolate bar for 28p, a bun for 7p and a cream cake for 22p.
(a) How much did she spend altogether?
(b) What change did she have from £1?

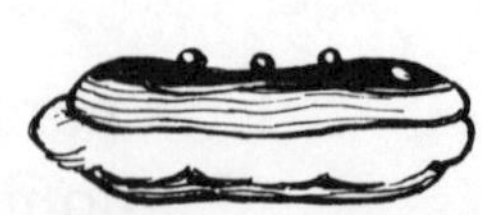

2. Mr and Mrs Bunter and their son, Billy, went up to London. The fare was £4 36p each. Billy's fare was £2 18p.
(a) What was the total fare for Mr and Mrs Bunter?
(b) How much was the fare for all three?
(c) How much change did Mr Bunter have from £20?

C

From £1 take:

1. 1p **2.** 2p **3.** 5p **4.** 10p **5.** 50p

D Work out:

1.	**2.**	**3.**	**4.**	**5.**
£	£	£	£	£
3·28	4·56	2·29	5·46	3·27
− 1·14	− 1·32	− 1·13	− 0·29	− 1·35

6.	**7.**	**8.**	**9.**	**10.**
£	£	£	£	£
6·26	5·24	3·22	7·09	5·07
− 1·48	− 0·85	− 0·17	− 2·46	− 0·24

11.	**12.**	**13.**	**14.**	**15.**
£	£	£	£	£
6·04	4·62	5·28	7·42	6·05
− 1·19	− 2·79	− 1·50	− 0·85	− 2·47

16.	**17.**	**18.**	**19.**	**20.**
£	£	£	£	£
4·62	10·81	5·58	8·07	7·57
− 3·79	− 1·99	− 0·75	− 1·90	− 6·68

21. £5·16 − £1·77

22. £10·02 − £4·56

23. £7·20 − £0·46

24. £3·84 − £1·93

E

1. The cost of 2 books was £8·75. One cost £3·99.
How much did the other cost?

2. A radio marked £15·15 was reduced to £12·75 in a sale.
How much was saved by buying at the lower price?

3. Anne saved £3·85. Her mother gave her £0·75.
(a) How much did she have then?
(b) How much more would she need to have saved £10?

4. Find the difference between £5·65 and £6·56.

Multiplying and dividing money

A Work out:

1.	**2.**	**3.**	**4.**	**5.**
p 2 × 4	p 3 × 6	p 4 × 5	p 6 × 7	p 8 × 9
6.	**7.**	**8.**	**9.**	**10.**
p 11 × 6	p 12 × 7	p 10 × 5	p 9 × 7	p 10 × 8
11.	**12.**	**13.**	**14.**	**15.**
£ p 10 × 10	£ p 15 × 7	£ p 16 × 8	£ p 46 × 9	£ p 54 × 5
16.	**17.**	**18.**	**19.**	**20.**
£ p 1 20 × 5	£ p 2 50 × 2	£ p 1 55 × 2	£ p 1 67 × 4	£ p 4 26 × 6

B Copy and complete these bills.

1. £ p

5 batteries at 64p each cost
3 light bulbs at 28p each cost
3 metres of flex at 19p a metre cost
TOTAL

2. £ p

2 blouses at £5 40p each cost
4 handkerchiefs at 32p each cost
3 pairs of tights at 75p a pair cost
TOTAL

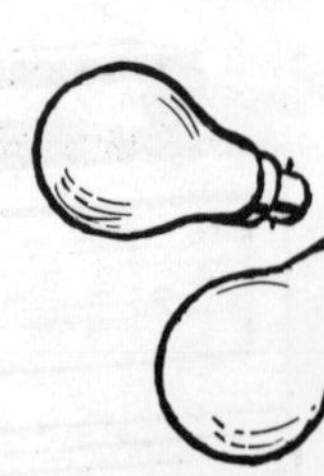

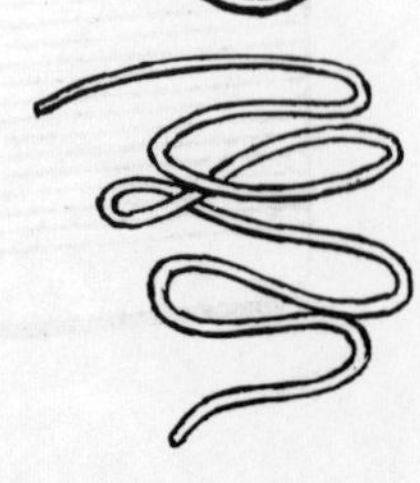

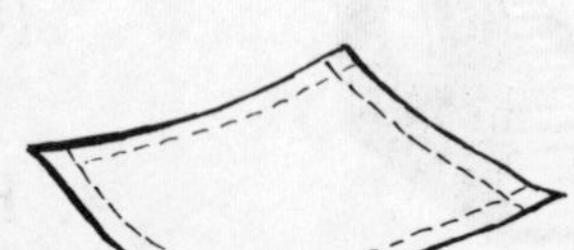

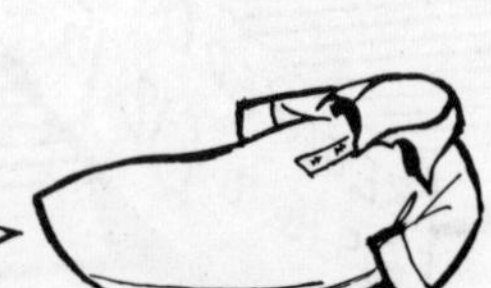

C Find the value of:

1. $\begin{array}{r} \text{p} \\ 8\overline{)32} \end{array}$ **2.** $\begin{array}{r} \text{p} \\ 9\overline{)36} \end{array}$ **3.** $\begin{array}{r} \text{p} \\ 8\overline{)40} \end{array}$ **4.** $\begin{array}{r} \text{p} \\ 8\overline{)64} \end{array}$ **5.** $\begin{array}{r} \text{p} \\ 3\overline{)18} \end{array}$

6. $\begin{array}{r} \text{p} \\ 6\overline{)24} \end{array}$ **7.** $\begin{array}{r} \text{p} \\ 7\overline{)56} \end{array}$ **8.** $\begin{array}{r} \text{p} \\ 5\overline{)25} \end{array}$ **9.** $\begin{array}{r} \text{p} \\ 9\overline{)72} \end{array}$ **10.** $\begin{array}{r} \text{p} \\ 6\overline{)36} \end{array}$

11. $\frac{45\text{p}}{5}$ **12.** $\frac{80\text{p}}{10}$ **13.** $\frac{54\text{p}}{6}$ **14.** $\frac{81\text{p}}{9}$ **15.** $\frac{63\text{p}}{7}$

16. 28p ÷ 4 **17.** 12p ÷ 6 **18.** 35p ÷ 7 **19.** 48p ÷ 6
20. 63p ÷ 9 **21.** 90p ÷ 10 **22.** 21p ÷ 3 **23.** 42p ÷ 7

D

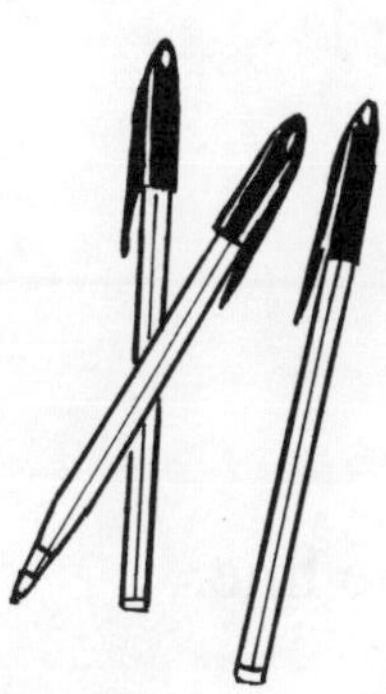

1. Diane bought a packet of 8 biros costing 72p.
(a) How much did each biro cost?
(b) Work out the cost of 2 biros, 5 biros, 7 biros, 10 biros.

2. Martin bought 2 cream buns for 15p each and 3 chocolate eclairs for 20p each.
(a) Work out the cost of the cream buns.
(b) Work out the cost of the chocolate eclairs.
(c) Find the total cost of the cakes.
(d) Find the average cost of each of the 5 cakes.

3. Mrs Black ordered 8 bottles of wine, together costing £32.
(a) How much was this for each bottle?
(b) How much would it cost for 2 bottles, 5 bottles, 9 bottles, 10 bottles?
(c) How many bottles would she have bought if she had paid £24?

4. Sue bought 8 balls of wool at 75p each.
(a) How much did she pay altogether?
(b) How many balls did she buy for £3?
(c) How many balls did she buy for £1·50?

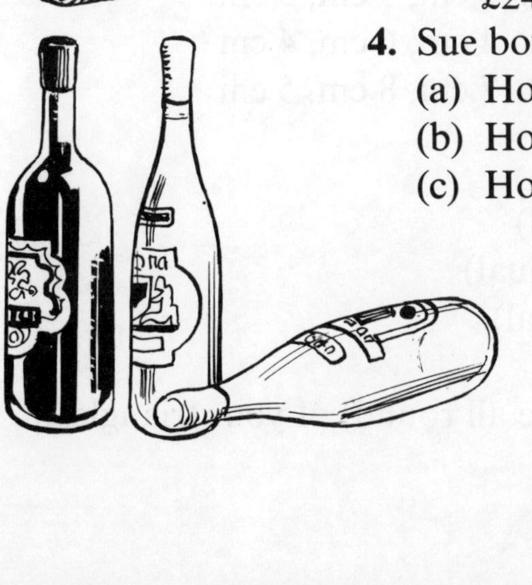

Triangles

There are some triangles which are special.
(i) Those with 2 sides equal are called ISOSCELES.
(ii) Those with all 3 sides equal are called EQUILATERAL.
SCALENE triangles have no sides equal to each other.

A You will need a pair of compasses to draw these triangles.
The first has sides 5 cm, 4 cm, 3 cm long.

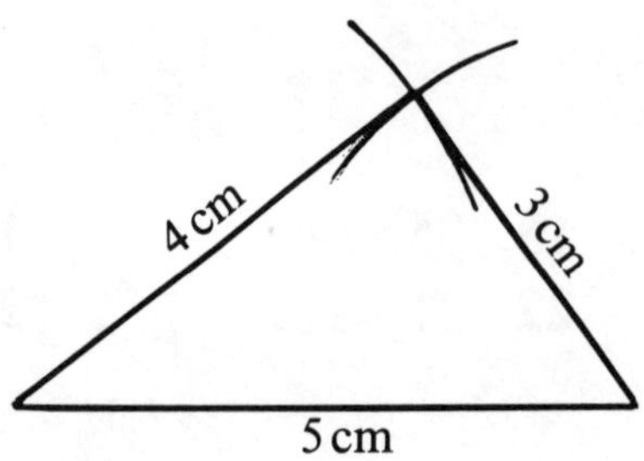

(i) Draw a line 5 cm long.
(ii) Open your compasses 4 cm wide.
(iii) Put the compass point on one end of the line.
(iv) With your compasses draw an ARC.
(v) Now open your compasses 3 cm wide.
(vi) Put your compass point on the other end of the line.
(vii) Draw another ARC with your compasses.
(viii) From where your two ARCS cross, draw lines to each end of the 5 cm line.

Now draw triangles with sides:

1. 6 cm, 4 cm, 3 cm
2. 7 cm, 5 cm, 5 cm
3. 2 cm, 3 cm, 4 cm
4. 5 cm, 5 cm, 5 cm
5. 5 cm, 12 cm, 13 cm
6. 4 cm, 6 cm, 4 cm
7. 5 cm, 6 cm, 6 cm
8. 7 cm, 8 cm, 5 cm

Write on your triangles if they are:
ISOSCELES (2 sides equal)
or EQUILATERAL (all 3 sides equal)
or SCALENE (no sides equal).

B Work out the PERIMETERS (distance all round) of your triangles.

C You can *fold* a RIGHT ANGLED triangle.
Take a sheet of paper—any piece.
(i) Fold the piece of paper.
(ii) Fold it again (use the diagram).

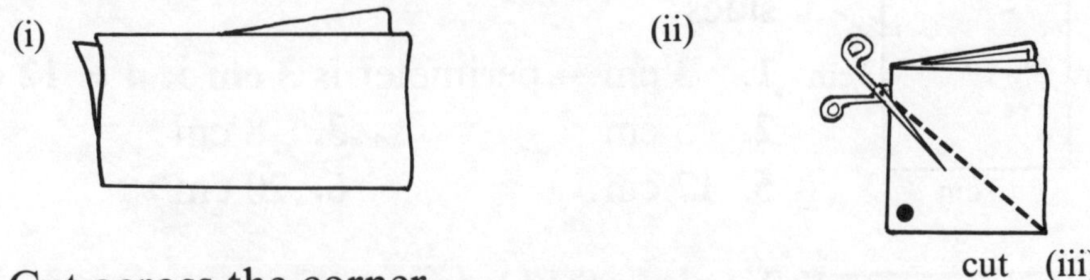

(iii) Cut across the corner.
This shape is a RIGHT ANGLED triangle.
The corner marked with a dot is a RIGHT ANGLE.
Which of the triangles you drew from page 32 was right angled?

D

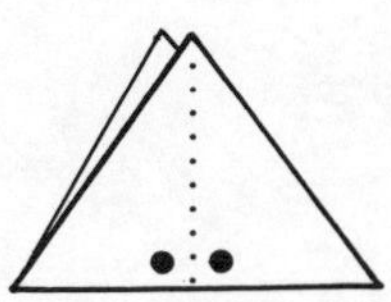

Open up your paper one fold to make another triangle.
Keep this.
This is an ISOSCELES triangle—2 sides have the same length.

Which of these triangles are ISOSCELES? Their sides are:

1. 3 cm, 3 cm, 4 cm
2. 5 cm, 4 cm, 3 cm
3. 5 cm, 12 cm, 13 cm
4. 5 cm, 4 cm, 5 cm
5. 3 cm, 2 cm, 2 cm
6. 8 cm, 7 cm, 6 cm

E A triangle with 3 sides all the same length is an EQUILATERAL triangle.
To fold an EQUILATERAL triangle take a sheet of paper and follow these instructions.

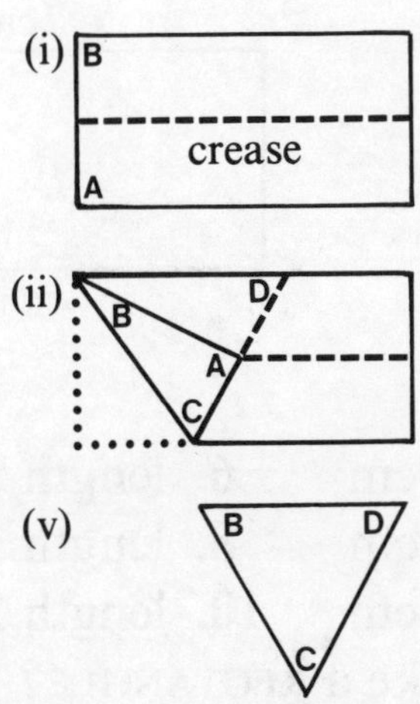

(i) Fold the paper in half lengthwise.
(ii) Fold corner A up to the crease.

(iii) Now fold along the line C A D

(iv) The triangle made by the corners B, D, C is EQUILATERAL.
(v) Cut out the triangle and measure its sides carefully.

Quadrilaterals

A Any shape with 4 straight sides is a QUADRILATERAL.
A SQUARE is a QUADRILATERAL.

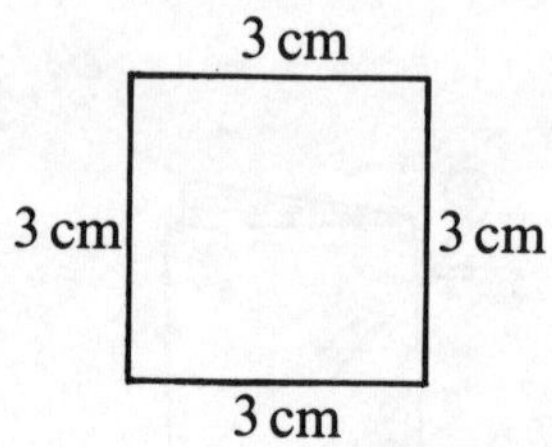

Work out the PERIMETERS (distance all round) of SQUARES with sides:

1. 3 cm → perimeter is **3 cm × 4 = 12 cm.**
2. 5 cm **3.** 8 cm **4.** 15 cm
5. 17 cm **6.** 20 cm **7.** 24 cm

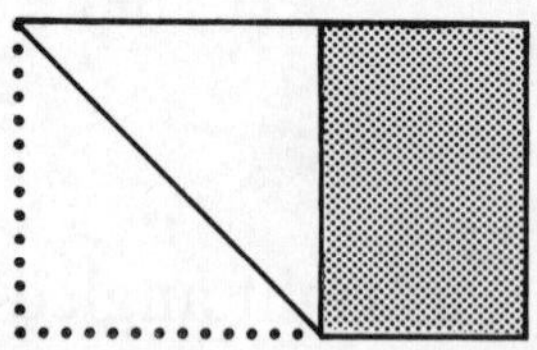

You can fold a SQUARE from a sheet of paper.
(i) Fold one corner to touch the opposite side.
(ii) Cut off the shaded part.
(iii) Open out the folded part. You have a SQUARE.

The fold is a DIAGONAL of the SQUARE.
Fold the other DIAGONAL.

B

The part of the paper shaded in the diagram is a RECTANGLE.
It has 2 equal long sides and 2 equal short sides.

Find the PERIMETER of these RECTANGLES:

1.
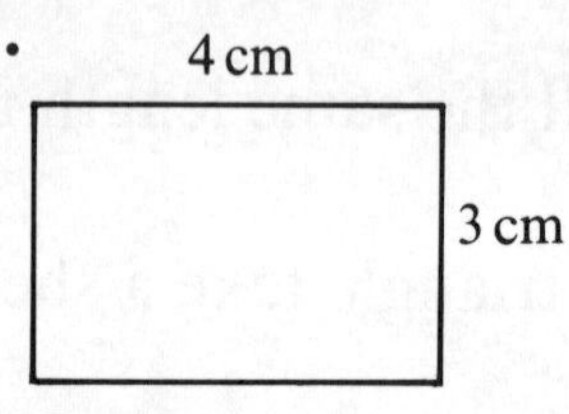

2.
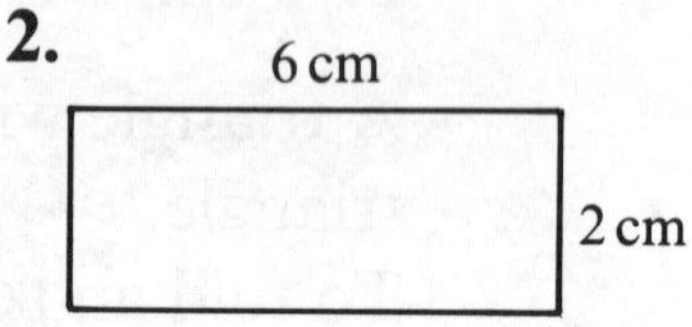

3.
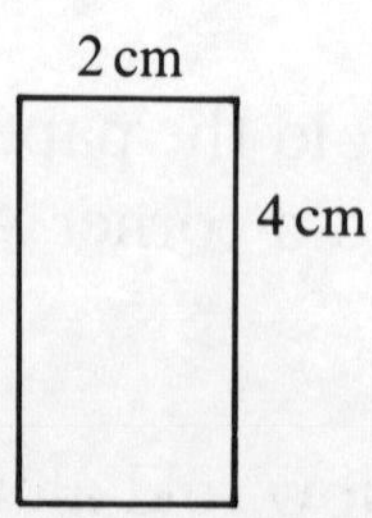

4.
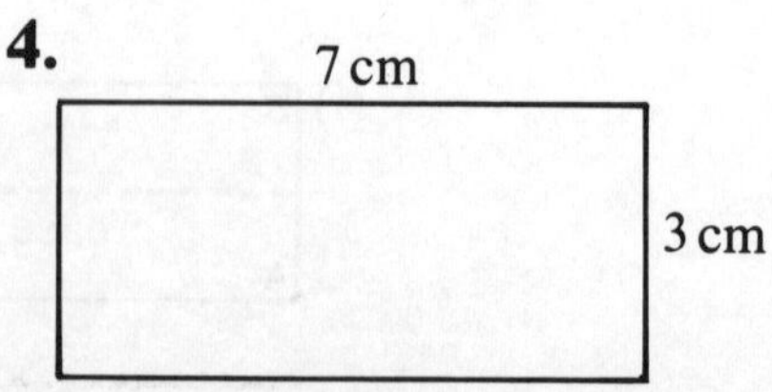

5. length 10 cm, width 7 cm **6.** length 12 cm, width 10 cm
7. length 17 cm, width 14 cm **8.** length 24 cm, width 13 cm
9. length 25 cm, width 15 cm **10.** length 28 cm, width 9 cm

In what way is a SQUARE like a RECTANGLE?

C

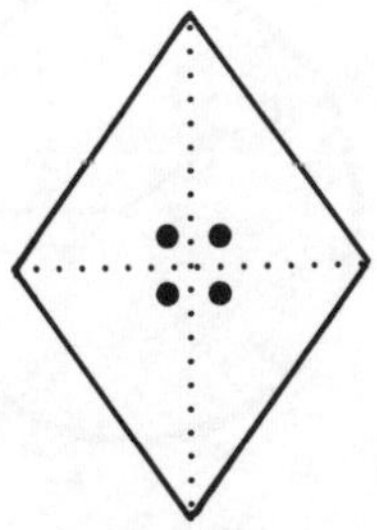

rhombus

Did you keep the ISOSCELES triangle you folded?
Now unfold it completely.
This new diamond shape is called a RHOMBUS.

1. How is a RHOMBUS different from a SQUARE?
2. What do you notice about the angles where the DIAGONALS meet?
3. Work out the PERIMETER of a RHOMBUS with sides:
(a) 2 cm (b) 6 cm (c) 7 cm (d) 9 cm
(e) 12 cm (f) 18 cm (g) 19 cm (h) 23 cm.

D

(i) Fold a piece of paper (any shape of paper).
(ii) Make another fold, but *not* along the previous one.
(iii) Cut as in the diagram.
(iv) Unfold the piece cut off.
This shape is called a KITE.

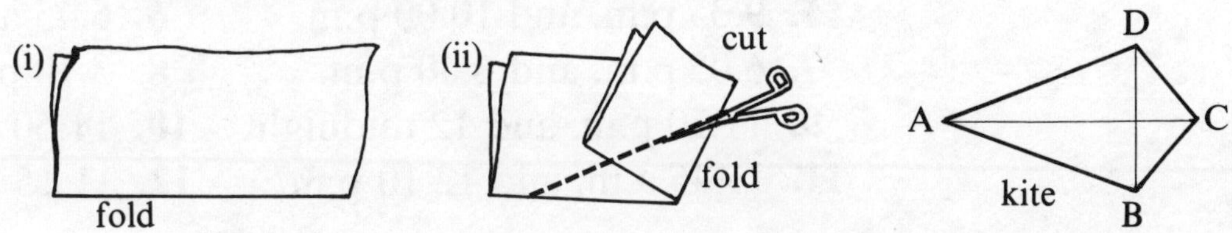

1. How is a KITE different from a RHOMBUS?
2. *Draw* the diagonal from A to C, along the fold.
Draw the other diagonal, the straight line joining B to D.
What can you say about the angles where the DIAGONALS cross?
3. Work out the PERIMETER of a KITE with sides:
(a) 3 cm and 4 cm (b) 3 cm and 6 cm
(c) 2 cm and 5 cm (d) 7 cm and 9 cm
(e) 5 cm and 9 cm (f) 12 cm and 10 cm.

E

parallelogram

A PARALLELOGRAM has its *opposite* sides equal and also its opposite sides PARALLEL (they never meet).

1. How does a KITE differ from a PARALLELOGRAM?
2. Work out the PERIMETER of a PARALLELOGRAM with sides:
(a) 5 cm and 7 cm (b) 12 cm and 9 cm
(c) 15 cm and 17 cm.
Make a sketch in each case.

Time

A

1. Write the times shown on each clock first in words, then in figures.
2. Record the *correct* time if each clock were 5 minutes fast.
3. Record the *correct* time if each clock were 10 minutes slow.

B

How many minutes are there between:

1. 2.20 a.m. and 2.40 a.m.
2. 1.30 p.m. and 1.55 p.m.
3. 1.45 p.m. and 2.05 p.m.
4. 12.55 p.m. and 1.05 p.m.
5. 9.35 p.m. and 10.00 p.m.
6. 6.45 a.m. and 7.30 a.m.
7. 6.05 p.m. and 6.40 p.m.
8. 3.10 p.m. and 4.00 p.m.
9. 11.50 p.m. and 12 midnight
10. 11.50 p.m. and 12.10 a.m.
11. 11.45 a.m. and 12.10 p.m.
12. 11.15 a.m. and NOON?

C

Write in figures:

1. 1 minute to midnight → **11.59 p.m.**
2. 5 minutes to noon
3. 10 minutes after midnight
4. 1 minute after noon
5. a quarter of an hour to midnight
6. half an hour to noon

D

There are 60 minutes in 1 hour.
How many minutes are there in:

1. $\frac{1}{2}$ an hour
2. $\frac{1}{4}$ hour
3. $\frac{3}{4}$ hour
4. $\frac{1}{3}$ hour
5. $\frac{2}{3}$ hour
6. $\frac{1}{6}$ hour
7. $\frac{5}{6}$ hour
8. 2 hours
9. $1\frac{1}{2}$ hours
10. 3 hours
11. 4 hours
12. 5 hours?

E

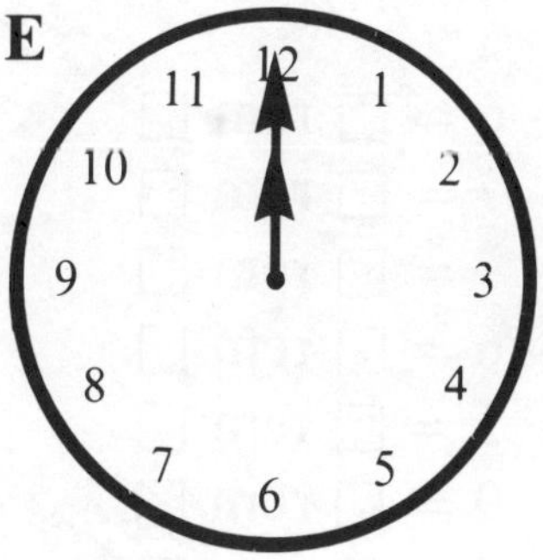

The clock is showing the time as noon.
Write the new time when it moves on:

1. 10 minutes **2.** 20 minutes **3.** 30 minutes
4. 50 minutes **5.** 60 minutes **6.** 65 minutes
7. 75 minutes **8.** 90 minutes **9.** 100 minutes
10. 110 minutes **11.** 120 minutes **12.** 2 hours.

F

There are 60 minutes in 1 hour.
Change to minutes:

1. 1 hour 10 minutes **2.** 1 hour 25 minutes
3. 1 hour 30 minutes **4.** 1 hour 45 minutes
5. 2 hours 5 minutes **6.** 2 hours 20 minutes
7. 3 hours 15 minutes **8.** 3 hours 20 minutes
9. 4 hours 10 minutes **10.** 5 hours 50 minutes.

G

Change to hours and minutes:

1. 70 minutes **2.** 90 minutes **3.** 100 minutes
4. 125 minutes **5.** 150 minutes **6.** 185 minutes

H

There are 60 seconds in 1 minute.
How many seconds are there in:

1. $\frac{1}{2}$ minute **2.** $\frac{1}{4}$ minute **3.** $\frac{3}{4}$ minute **4.** $\frac{1}{3}$ minute
5. $\frac{2}{3}$ minute **6.** $1\frac{1}{2}$ minutes **7.** 2 minutes **8.** 3 minutes?

I

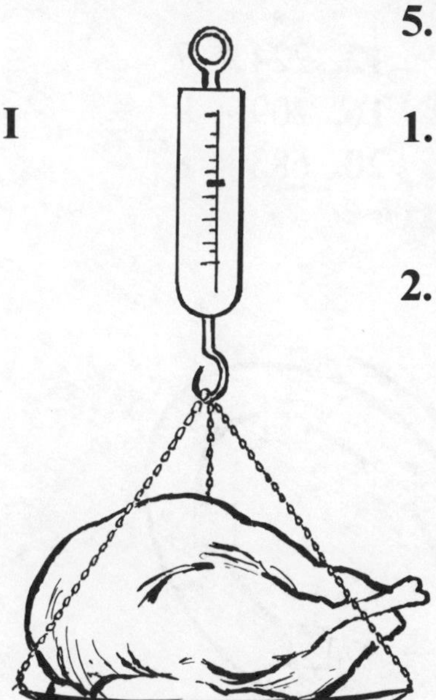

1. A patient's pulse beats 18 times in 15 seconds.
How many beats would take: (a) 30 seconds (b) 45 seconds (c) 1 minute?

2. Cooking time for a turkey is 40 minutes for each kilogram plus 20 minutes. Work out the cooking time (in minutes) for a turkey weighing:
(a) 2 kg (b) 3 kg (c) 4 kg (d) 6 kg.

Dividing

A Time yourself doing these.

1. 17 ÷ 2 = **8 rem 1** **2.** 14 ÷ 3 = **4 rem 2** **3.** 19 ÷ 6 = □ rem □
4. 15 ÷ 4 = □ rem □ **5.** 9 ÷ 4 = □ rem □ **6.** 17 ÷ 7 = □ rem □
7. 25 ÷ 6 = □ rem □ **8.** 26 ÷ 3 = □ rem □ **9.** 29 ÷ 7 = □ rem □
10. 23 ÷ 5 = □ rem □ **11.** 27 ÷ 6 = □ rem □ **12.** 29 ÷ 6 = □ rem □
13. 33 ÷ 4 = □ rem □ **14.** 35 ÷ 8 = □ rem □ **15.** 32 ÷ 5 = □ rem □
16. 36 ÷ 7 = □ rem □ **17.** 39 ÷ 7 = □ rem □ **18.** 38 ÷ 9 = □ rem □
19. 42 ÷ 5 = □ rem □ **20.** 44 ÷ 8 = □ rem □ **21.** 50 ÷ 7 = □ rem □
22. 52 ÷ 8 = □ rem □ **23.** 55 ÷ 6 = □ rem □ **24.** 56 ÷ 9 = □ rem □

B *Examples:* **1.** 35 ÷ 2 **2.** 36 ÷ 4 **3.** 189 ÷ 6 **4.** 146 ÷ 7

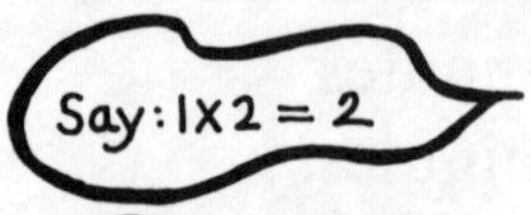

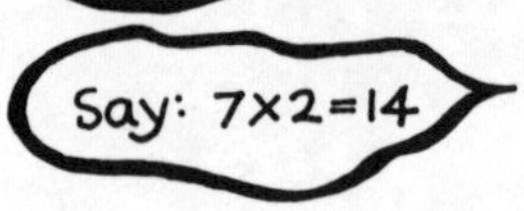

1. $\begin{array}{r} 17 \\ 2\overline{)35} \\ -2\downarrow \\ 15 \\ -14 \\ 1 \end{array}$ **2.** $\begin{array}{r} 36 \\ 4\overline{)145} \\ -12 \\ 25 \\ -24 \\ 1 \end{array}$ **3.** $\begin{array}{r} 31 \\ 6\overline{)189} \\ -18 \\ 09 \\ -6 \\ 3 \end{array}$ **4.** $\begin{array}{r} 20 \\ 7\overline{)146} \\ -14\downarrow \\ 06 \end{array}$

answer 17 rem 1 **answer 36 rem 1** **answer 31 rem 3** **answer 20 rem 6**

Work out:

1. $4\overline{)35}$ **2.** $2\overline{)69}$ **3.** $4\overline{)58}$ **4.** $3\overline{)73}$
5. $5\overline{)66}$ **6.** $4\overline{)70}$ **7.** $7\overline{)89}$ **8.** $6\overline{)79}$
9. 91 ÷ 6 **10.** 126 ÷ 5 **11.** 128 ÷ 4 **12.** 224 ÷ 7
13. 250 ÷ 3 **14.** 139 ÷ 8 **15.** 273 ÷ 9 **16.** 209 ÷ 9
17. 283 ÷ 7 **18.** 241 ÷ 6 **19.** 424 ÷ 7 **20.** 683 ÷ 8

C Complete the arrow graphs.

1.

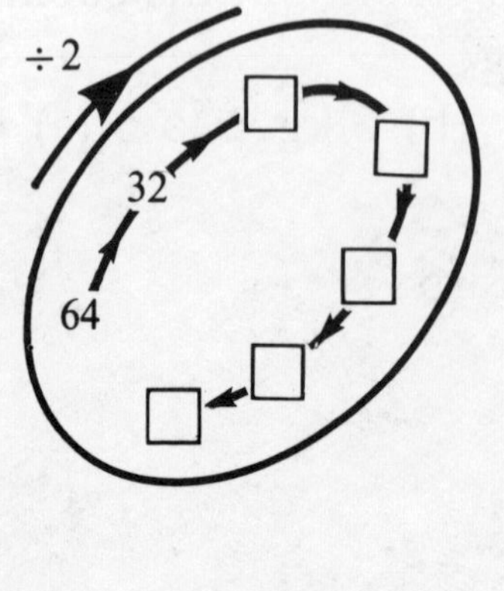

2.

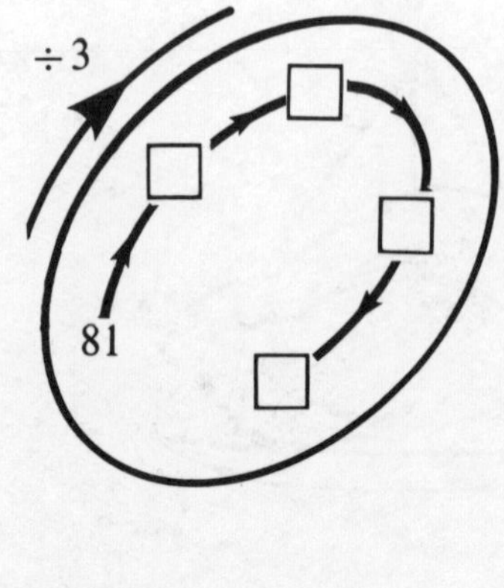

D

1. There are 7 days in a week. There are 365 days in 1 year.
(a) How many weeks are there in a year?
(b) How many days are left over?

2. The numbers of matches in 5 boxes were:
48, 54, 47, 52, 49.
(a) How many matches were there altogether?
(b) What was the average number in a box?

3. A cyclist travelled 217 km in 7 hours.
(a) How far did he go in one hour on average?
(b) What would this be for 2 hours? for 3 hours? for 5 hours?

E

Copy and complete the table for the number of theatre seats sold.

SEATS	COST OF TICKET	NUMBER OF TICKETS	TOTAL TAKINGS
upper circle	£3		£327
circle	£5		£405
front stalls	£7		£294
rear stalls	£4		£172
boxes		8	£96

Multiplying and dividing lengths

A Work out:

1.	**2.**	**3.**	**4.**	**5.**
cm	cm	cm	cm	cm
5	7	8	9	4
× 6	× 9	× 6	× 8	× 7

6.	**7.**	**8.**	**9.**	**10.**
cm	cm	cm	cm	cm
12	15	36	45	63
× 6	× 7	× 3	× 4	× 5

11.	**12.**	**13.**	**14.**	**15.**
m cm	m cm	m cm	m cm	m cm
43	24	1 21	2 56	1 77
× 5	× 7	× 4	× 4	× 5

16.	**17.**	**18.**	**19.**	**20.**
m	m	m	m	m
0·57	3·09	0·43	1·60	1·75
× 8	× 2	× 6	× 3	× 8

B

1. A swimming pool is 25 metres long. How far has a swimmer gone after:
(a) 2 lengths (b) 4 lengths (c) 7 lengths (d) 10 lengths?

2. Find the perimeters (distance all round) of squares with sides:
(a) 13 metres (b) 25 cm (c) 1 m 10 cm (d) 1·25 m.
Give your answers in metres.

3. A beach ball has a CIRCUMFERENCE (distance all round) of 85 cm.
When it rolls round once it will travel 85 cm.
How far will it travel after rolling round:
(a) twice (b) 3 times (c) 7 times (d) 10 times?
Give your answers in metres.

C Work out:

1. cm	**2.** cm	**3.** cm	**4.** cm	**5.** cm
$7\overline{)14}$	$8\overline{)56}$	$7\overline{)63}$	$9\overline{)81}$	$3\overline{)21}$
6. cm	**7.** cm	**8.** cm	**9.** cm	**10.** cm
$6\overline{)72}$	$5\overline{)60}$	$8\overline{)96}$	$5\overline{)75}$	$3\overline{)51}$
11. 45 m ÷ 3	**12.** 91 m ÷ 7	**13.** 80 m ÷ 10	**14.** 98 m ÷ 7	**15.** 108 m ÷ 9
16. $\frac{1{\cdot}44\text{ m}}{8}$	**17.** $\frac{1{\cdot}25\text{ m}}{5}$	**18.** $\frac{1{\cdot}60\text{ m}}{4}$	**19.** $\frac{24{\cdot}60\text{ m}}{5}$	**20.** $\frac{1{\cdot}19\text{ m}}{7}$
21. m cm	**22.** m cm	**23.** m cm	**24.** m cm	**25.** m cm
$7\overline{)1\ \ 26}$	$9\overline{)1\ \ 35}$	$5\overline{)4\ \ 25}$	$6\overline{)3\ \ 36}$	$8\overline{)4\ \ 56}$

D **1.** A strip of sellotape 56 cm long is cut into 7 equal pieces.
(a) How long is one piece?
(b) How long are 2 pieces? 3 pieces? 5 pieces?

2. 9 pennies in a line touching each other stretch 18 cm.
(a) How wide is a 1p coin?
(b) How far do 4 coins stretch? 6 coins? 20 coins?

3. A piece of paper 30 cm wide was cut into strips.
How many strips are there if each one were
(a) 3 cm wide (b) 5 cm wide (c) 6 cm wide (d) 10 cm wide?

4. Work out the length of the side of a square whose PERIMETER is:
(a) 16 cm (b) 48 cm (c) 72 cm (d) 104 cm.

5. Four vehicles each travel 24 metres. The circumferences of the wheels are:
(a) 4 m (b) 3 m (c) 2 m (d) 1 m.
How many times do the wheels of each turn?
Suggest what vehicles they might be.

Weight – grams

A A set of weights is made from one each of these.

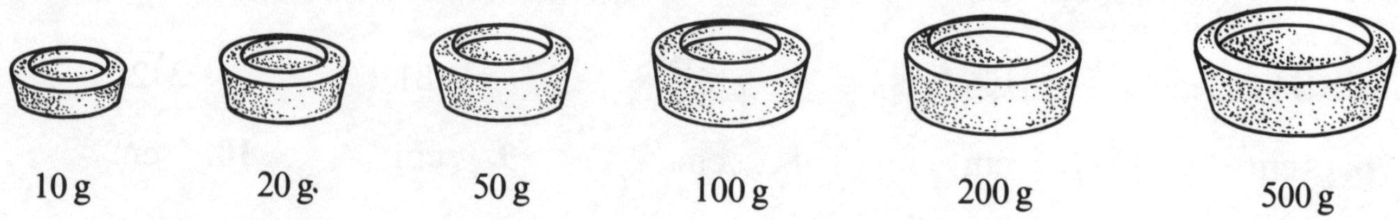

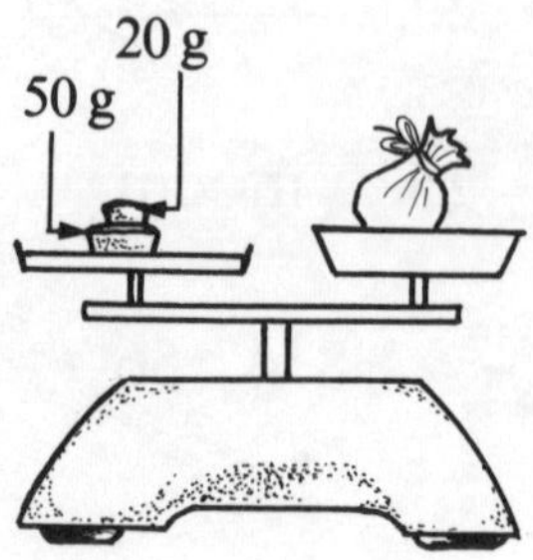

To weigh 70 grams (70 g) use 50 g + 20 g → 70 g.
Which weights would you use to weigh:

1. 30 g **2.** 80 g **3.** 130 g **4.** 120 g
5. 60 g **6.** 160 g **7.** 170 g **8.** 210 g
9. 230 g **10.** 260 g **11.** 310 g **12.** 330 g
13. 360 g **14.** 370 g **15.** 380 g **16.** 510 g
17. 520 g **18.** 530 g **19.** 560 g **20.** 750 g
21. What is the most you can weigh with this set of weights?

B

A little more thought is needed to weigh some amounts.
To weigh 40 g put 50 g in one pan, and 10 g in the pan with the 40 g amount.

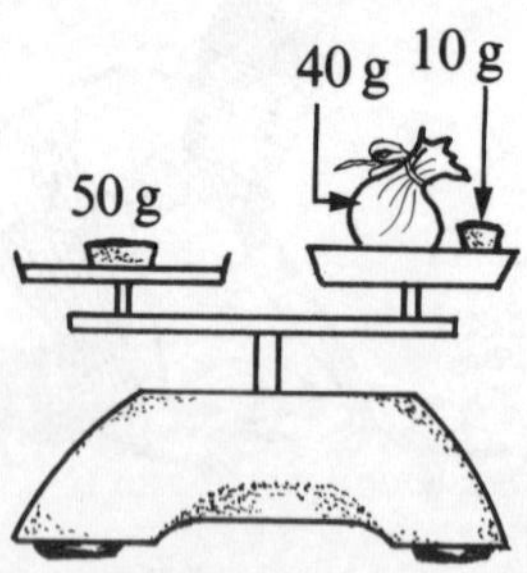

50 g – 10 g = 40 g
To weigh 190 g use 200 g – 10 g.

Which weights would you use for these?

1. 90 g **2.** 140 g **3.** 240 g **4.** 290 g **5.** 340 g
6. 400 g **7.** 490 g **8.** 450 g **9.** 480 g **10.** 540 g
11. 590 g **12.** 390 g **13.** 440 g **14.** 640 g **15.** 790 g
16. Think of a way to make a 5 g weight using Plasticine.

C Copy and complete the table to show the weights you would use for the following:

WEIGHT	500 g	200 g	100 g	50 g	20 g	10 g
70 g	**0**	**0**	**0**	**1**	**1**	**0**
110 g						
130 g						
180 g						
220 g						
270 g						
380 g						
620 g						
760 g						
870 g						

D Find the total weight of the ingredients in these recipes.

1. Short crust pastry	
	flour 250 g
	lard or margarine 125 g
	salt 5 g

2. Bread pudding	
	bread 250 g
	sugar 50 g
	suet 50 g
	dried fruit 100 g
	spice 5 g
	1 egg 60 g

3. Work out the quantities of each of the ingredients in these recipes if twice the amount is required.

Adding and subtracting weights

A Work out:

1.	**2.**	**3.**	**4.**	**5.**
g 35 + 56	g 64 + 37	g 59 + 42	g 84 + 16	g 29 + 79
6.	**7.**	**8.**	**9.**	**10.**
g 87 + 46	g 94 + 119	g 37 + 184	g 198 + 52	g 276 + 114
11.	**12.**	**13.**	**14.**	**15.**
g 264 + 117	g 273 + 227	g 324 + 256	g 406 + 294	g 564 + 379

16. 527 g + 75 g + 28 g **17.** 510 g + 10 g + 85 g **18.** 452 g + 225 g
19. 9 g + 334 g + 74 g **20.** 48 g + 512 g + 5 g **21.** 625 g + 20 g

B

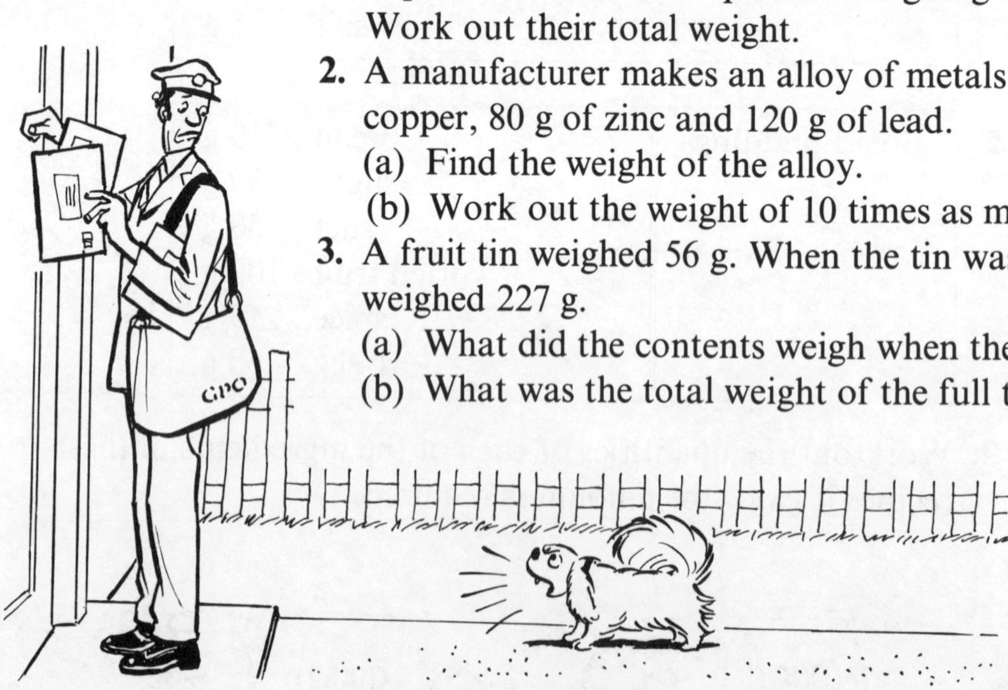

1. A postman delivered 3 packets weighing 125 g, 250 g, 50 g. Work out their total weight.
2. A manufacturer makes an alloy of metals using 100 g of copper, 80 g of zinc and 120 g of lead.
   (a) Find the weight of the alloy.
   (b) Work out the weight of 10 times as much alloy.
3. A fruit tin weighed 56 g. When the tin was half full, the fruit weighed 227 g.
   (a) What did the contents weigh when the tin was full?
   (b) What was the total weight of the full tin?

C Make a list of things whose weight would be measured in grams (g).

D Work out:

1. g	**2.** g	**3.** g	**4.** g	**5.** g
76	58	127	123	172
− 35	− 49	− 46	− 87	− 158

6. g	**7.** g	**8.** g	**9.** g	**10.** g
215	250	453	462	660
− 125	− 175	− 256	− 415	− 453

11. g	**12.** g	**13.** g	**14.** g	**15.** g
503	750	800	900	400
− 225	− 295	− 255	− 367	− 125

16. 453 g − 375 g　　**17.** 402 g − 175 g　　**18.** 326 g − 79 g
19. 575 g − 216 g　　**20.** 516 g − 9 g　　**21.** 453 g − 226 g
22. 729 g − 453 g　　**23.** 900 g − 52 g　　**24.** 750 g − 75 g

E

1. A tin of peas weighs 361 g. On the tin it reads 'Net weight 283 g', meaning the peas inside the tin weigh 283 g. Find the weight of the tin.

2. A jar half full of coffee weighed 340 g.
When it was full it weighed 390 g.
(a) What weight of coffee had been used?
(b) What weight of coffee had the jar contained? (net weight)
(c) How much did the jar weigh?

3. John sent Paula a box of chocolates. The packet weighed 500 g.
The chocolates weighed 453 g.
How much did the packaging weigh?

Multiplying and dividing grams

A Work out:

1.	**2.**	**3.**	**4.**	**5.**
g	g	g	g	g
27	25	19	46	37
× 3	× 4	× 5	× 6	× 6

6.	**7.**	**8.**	**9.**	**10.**
g	g	g	g	g
67	78	49	37	80
× 9	× 7	× 7	× 9	× 7

11.	**12.**	**13.**	**14.**	**15.**
g	g	g	g	g
125	125	250	125	250
× 4	× 2	× 3	× 6	× 2

16.	**17.**	**18.**	**19.**	**20.**
g	g	g	g	g
225	112	375	128	176
× 4	× 6	× 3	× 6	× 4

B

1. Each letter in a bundle weighs 28 g.
Find the weight of:
(a) 2 letters (b) 5 letters (c) 6 letters (d) 10 letters.

2. Each chocolate in a box weighs 7 g.
Find the net weight of the chocolates with:
(a) 10 in the box (b) 15 in the box (c) 25 in the box
(d) 36 in the box (e) 48 in the box (f) 60 in the box.

3. A torch battery weighs 56 g. Find the weight of:
(a) 4 batteries (b) 6 batteries (c) 8 batteries
(d) 9 batteries (e) 10 batteries (f) 20 batteries.

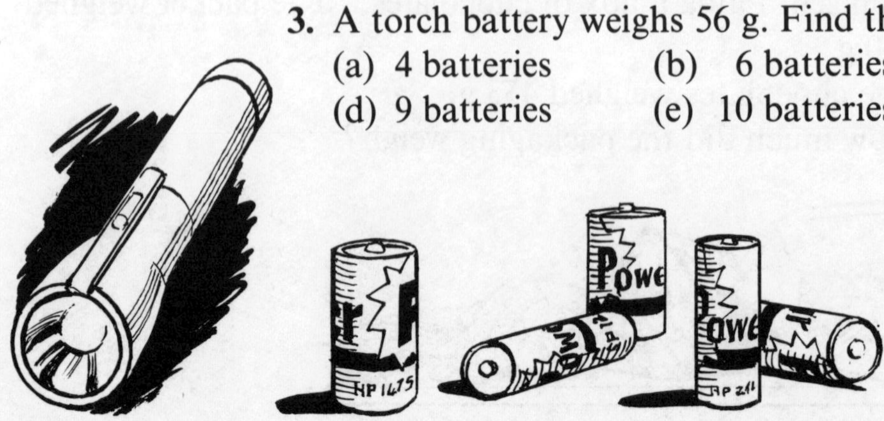

C *Examples:* **1.** 15 g ÷ 5 **2.** 150 g ÷ 5 **3.** 154 g ÷ 5

1.
```
   g
   3
5)15
 -15
   0
```
→ **2.**
```
    g
   30
5)150
 -15↓
    0
```
→ **3.**
```
    g
  30 4/5
5)154
 -15↓
    4
```

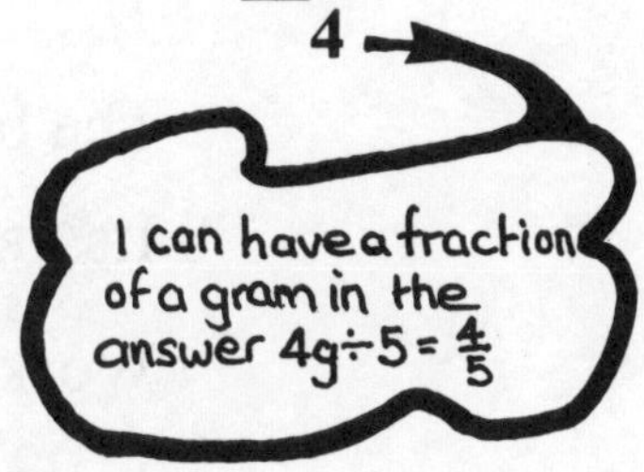

Work out:

1. g 3)15 → **2.** g 3)150 **3.** g 6)42 → **4.** g 6)420 **5.** g 4)24

6. g 4)240 **7.** g 8)48 **8.** g 8)480 **9.** g 4)36 **10.** g 4)360

11. g 8)320 **12.** g 7)210 **13.** g 5)400 **14.** g 9)630 **15.** g 8)640

16. g 5)304 **17.** g 6)421 **18.** g 4)243 **19.** g 8)485 **20.** g 5)402

D

1. A bar of chocolate weighing 454 g was shared equally among 9 children.
(a) How much did each have?
(b) How much would each have if there were only 5 children?

2. A packet of 10 felt-tipped pens weighs 90 g.
(a) Find the weight of one pen.
(b) Find the weight of 4 pens, 7 pens, 9 pens.

3. The dry ingredients for chicken pie for 10 people are:
200 g ham, 50 g butter, 400 g cooked chicken, 150 g mushrooms, 500 g short crust pastry.
Work out the ingredients needed to serve:
(a) 5 people (b) 2 people.

Area of a rectangle

A AREA is measured using squares to cover the surface.

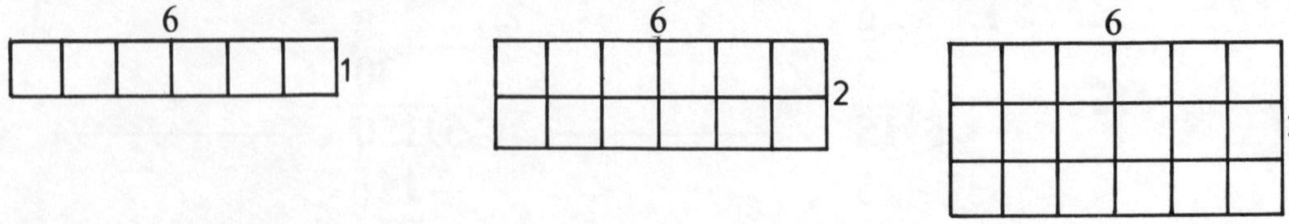

The first RECTANGLE is covered by one row of 6 squares.

Its AREA is 6 SQUARE UNITS.

Work out the AREA of RECTANGLES with 6 squares in a row, and:

1. 2 deep **2.** 3 deep **3.** 4 deep
4. 7 deep **5.** 9 deep **6.** 10 deep
7. 20 deep **8.** 25 deep **9.** 45 deep.

B The usual unit for measuring a small AREA is a SQUARE CENTIMETRE (cm^2).

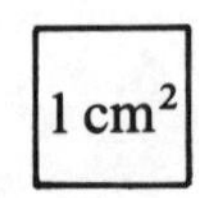

These RECTANGLES cover 12 cm^2 (12 square centimetres).

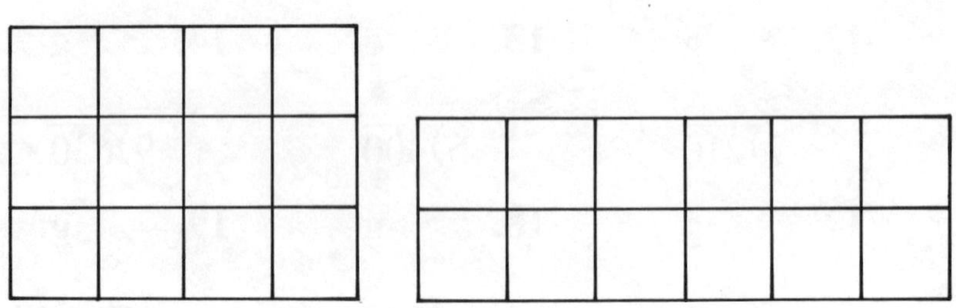

One has 3 rows of 4 squares → 12 cm^2.
The other has 2 rows of 6 squares → 12 cm^2.

Draw as many different RECTANGLES as you can with these AREAS.

1. 12 cm^2 **2.** 18 cm^2 **3.** 20 cm^2 **4.** 24 cm^2
5. 28 cm^2 **6.** 30 cm^2 **7.** 36 cm^2 **8.** 42 cm^2

It may help to use centimetre squared paper.

C What could be the length and breadth of 2 RECTANGLES, each of whose AREA is:

1. 48 cm^2 **2.** 54 cm^2 **3.** 56 cm^2 **4.** 63 cm^2?

D Draw the following on centimetre squared paper.
Work out their AREAS in SQUARE CENTIMETRES (cm^2).

1. 5 cm by 4 cm	**2.** 3 cm by 6 cm	**3.** 8 cm by 4 cm
4. 7 cm by 5 cm	**5.** 6 cm by 8 cm	**6.** 4 cm by 9 cm
7. 6 cm by 6 cm	**8.** 12 cm by 3 cm	**9.** 18 cm by 2 cm
10. 2 cm by 8 cm	**11.** 10 cm by 3 cm	**12.** 6 cm by 3 cm

E Find the WIDTH of these RECTANGLES.

1. Area 14 cm^2, length 7 cm	**2.** Area 16 cm^2, length 8 cm
3. Area 42 cm^2, length 6 cm	**4.** Area 20 cm^2, length 5 cm
5. Area 48 cm^2, length 8 cm	**6.** Area 18 cm^2, length 6 cm
7. Area 54 cm^2, length 9 cm	**8.** Area 56 cm^2, length 8 cm

F Copy and complete the table for these RECTANGLES.

LENGTH (cm)	7		5		12	9	16	
BREADTH (cm)	5	7		6	7	8	4	2
AREA (cm^2)		63	55	54				28

Complete the statement:
The AREA of a RECTANGLE = L______ × B______.

G

1. A piece of card 8 cm wide has an AREA of 80 cm^2.
   (a) Find its length.
   (b) Find its PERIMETER (distance all round).

2. The PERIMETER of a RECTANGLE is 60 cm. It is 5 cm wide.
   (a) Find the distance half way round.
   (b) Work out its length.
   (c) Find the AREA of the RECTANGLE.

Area of squares

A

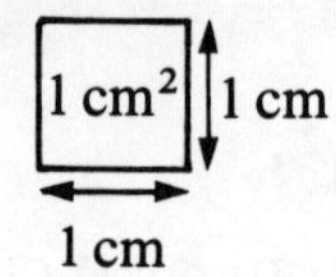

A 1 cm SQUARE is a square with sides 1 cm long.
Its AREA is 1 SQUARE CENTIMETRE (1 cm^2).

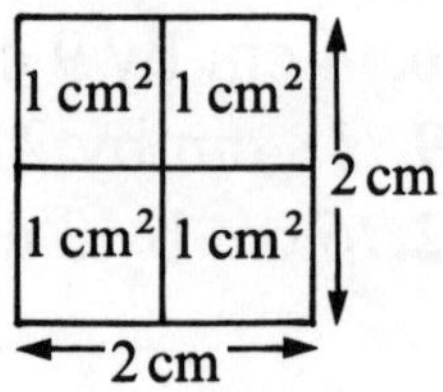

A 2 cm SQUARE has sides 2 cm long.
But its AREA is 4 SQUARE CENTIMETRES (4 cm^2).

Describe these SQUARES and then work out the AREA of each.

1.

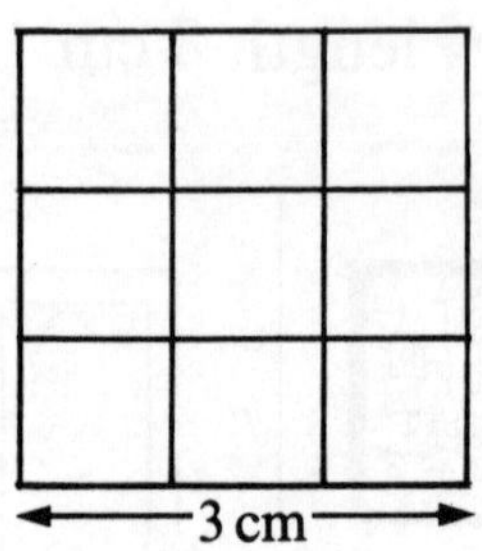

2.

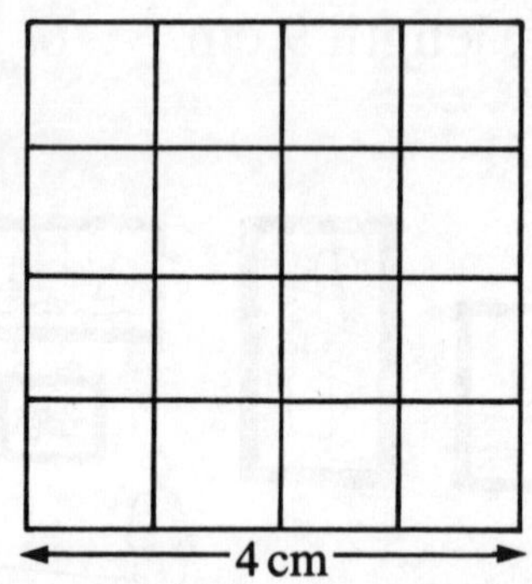

3.

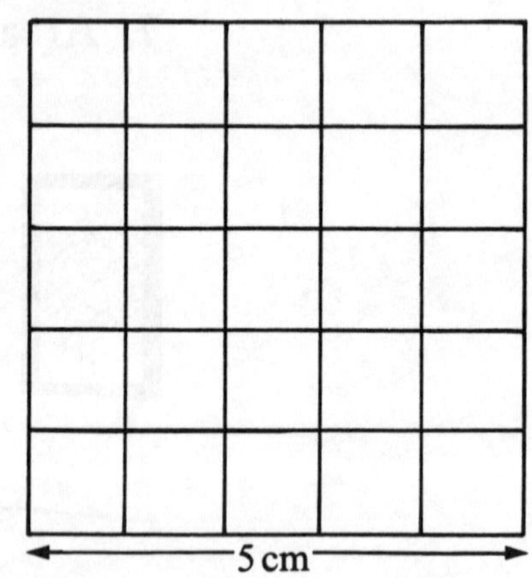

B

On centimetre squared paper draw these SQUARES.
Work out the AREA of each.

1. 6 cm square **2.** 7 cm square **3.** 8 cm square
4. 9 cm square **5.** 10 cm square **6.** 11 cm square

C

Look at the 2 cm SQUARE in Section A. Its AREA is 4 cm^2.
The difference in AREA between a 2 cm SQUARE and 2 cm^2 is $(4 - 2)\ cm^2 = 2\ cm^2$.

Work out the difference in AREA between:

1. a 3 cm square and 3 cm^2 **2.** a 4 cm square and 4 cm^2
3. a 5 cm square and 5 cm^2 **4.** a 6 cm square and 6 cm^2
5. a 7 cm square and 7 cm^2 **6.** a 9 cm square and 9 cm^2.

D

A 4 cm SQUARE has a side *twice* as big as a 2 cm SQUARE
but
the AREA of the 4 cm SQUARE is *four times* as big as a 2 cm SQUARE.

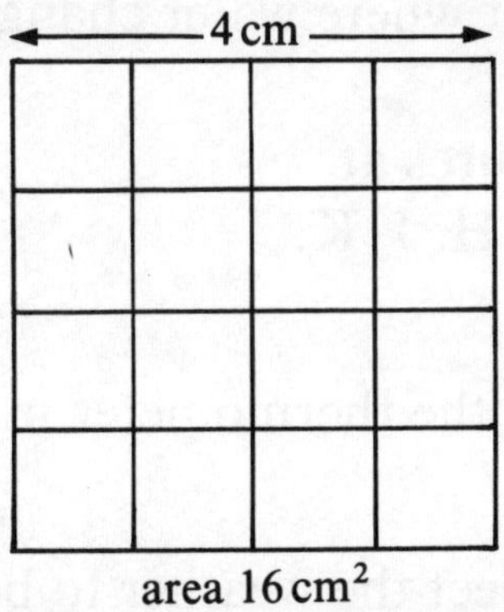

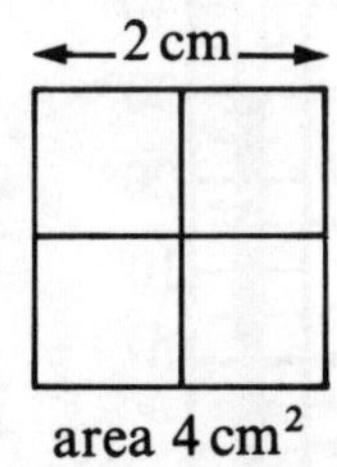

How many TIMES bigger is the AREA of

1. an 8 cm square than a 4 cm square?
2. a 9 cm square than a 3 cm square?
3. a 6 cm square than a 2 cm square?
4. a 10 cm square than a 5 cm square?
5. a 10 cm square than a 2 cm square?

E

Find the length of the side of the SQUARE with an AREA of:

1. 4 cm^2 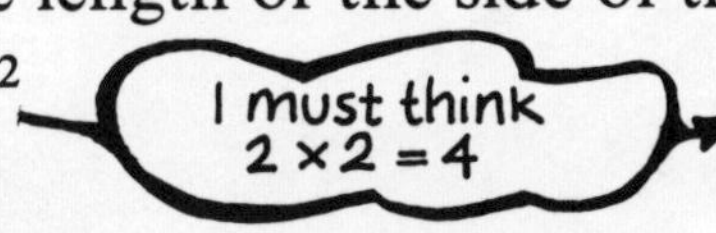

2 cm

2. 16 cm^2 **3.** 25 cm^2 **4.** 100 cm^2 **5.** 9 cm^2
6. 36 cm^2 **7.** 64 cm^2 **8.** 49 cm^2 **9.** 81 cm^2.

F

1. A square has a PERIMETER of 28 cm.
(a) Find the length of each side.
(b) Work out the AREA of the square.

2. A square has an AREA of 25 cm^2.
(a) Find the length of each side.
(b) Work out the PERIMETER of the square.
(c) Work out the PERIMETER of 2 such squares joined together like this:

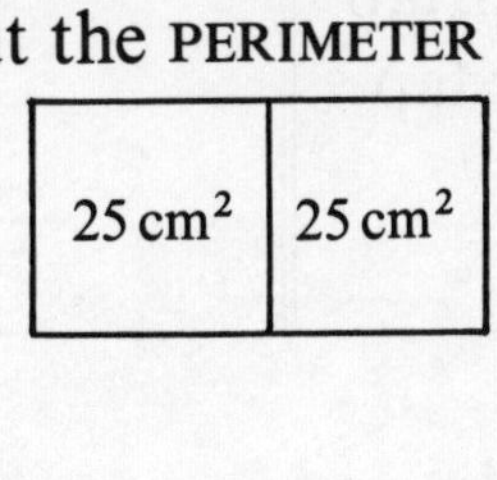

The thermometer

A

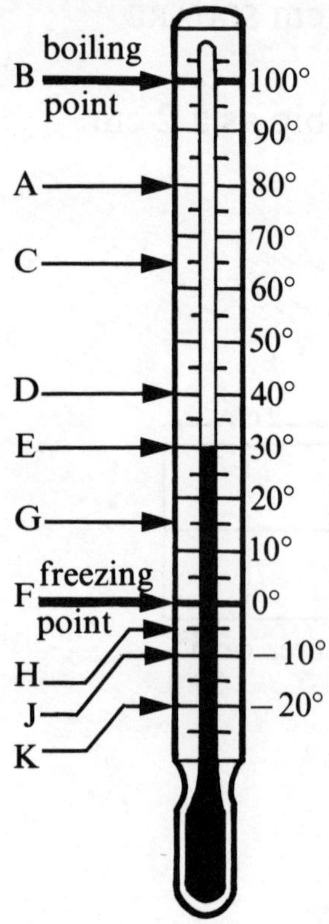

This is a CELSIUS thermometer.
Anders Celsius divided his thermometer into 100 divisions – 100 degrees – 100°
0° is the FREEZING point where water turns to ice.
100° is the BOILING point where water changes to steam.

1. Record the temperatures at A, B, C, D, E, F, G, H, J, K.
   Write **A → 80°**.
2. What temperature is the thermometer in the picture recording?
3. What would you expect the weather to be with this thermometer reading?
4. What would you expect the weather to be with the reading at H?
5. Find how near to D your blood temperature is.
6. What is the lowest temperature this thermometer will record?
7. What is the highest temperature you could record with this thermometer?

B

Use a Celsius thermometer to make these recordings.
Copy and complete the table.
Show the change in temperature from one reading to the next.

	TEMPERATURE	CHANGE IN TEMPERATURE
class room		—
corridor		
cold water tap		
hot water tap		
radiator		
my hand		

C

Look at the thermometer on the opposite page.
The rise in temperature from E → D is 10°: write **+10°**.
The fall in temperature from C → D is 10°: write **−10°**.
Record the RISE or FALL in temperature from:

1. D → B	**2.** B → D	**3.** F → E	**4.** E → J
5. J → H	**6.** H → C	**7.** C → A	**8.** A → F
9. F → B	**10.** B → K	**11.** K → G	**12.** G → A
13. A → K	**14.** K → E	**15.** E → H	**16.** H → F.

D

Find out:

1. what happens to the temperature between daytime and night-time.
2. what happens when a thermometer is moved from a sunny place into the shade.
3. what causes the liquid in the thermometer to rise or fall with the changing temperature.
4. what happens to the temperature between summer and winter.
5. what happens to your temperature if you have a fever.

E

SUBSTANCE	MELTING POINT
tin	232°C
copper	1083°C
butter	30°C
wax	50°C
iron	1530°C

Work out in degrees the difference in melting point between:

1. wax and butter
2. copper and tin
3. iron and copper
4. iron and tin
5. iron and butter.

F

Another thermometer is the FAHRENHEIT thermometer.
A rough method to change Celsius readings to Fahrenheit is:
double the Celsius and add 30°: 10°C → 10° × 2 + 30° → 50°F.
Change to Fahrenheit:

1. 8°C **2.** 12°C **3.** 20°C **4.** 30°C.

Fractions

A

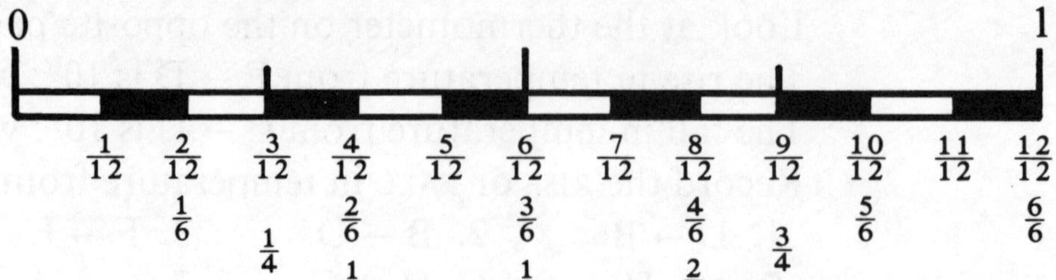

Use the twelfths number line to write, in lowest terms:

1. $\frac{6}{12}$ **2.** $\frac{2}{12}$ **3.** $\frac{4}{12}$ **4.** $\frac{9}{12}$ **5.** $\frac{8}{12}$ **6.** $\frac{10}{12}$ **7.** $\frac{12}{12}$.

B

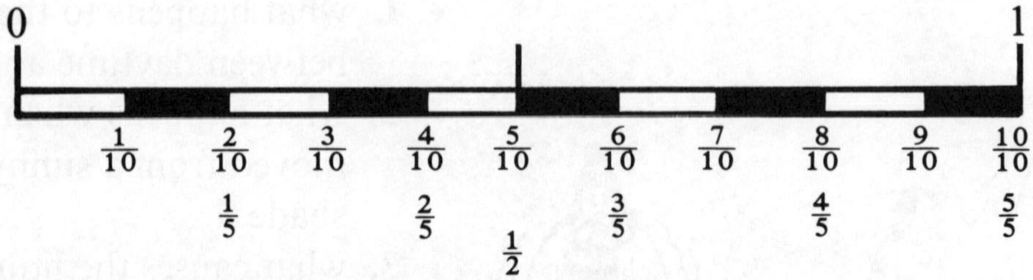

Use the tenths number line to write, in lowest terms:

1. $\frac{2}{10}$ **2.** $\frac{5}{10}$ **3.** $\frac{6}{10}$ **4.** $\frac{8}{10}$ **5.** $\frac{4}{10}$ **6.** $\frac{10}{10}$.

C

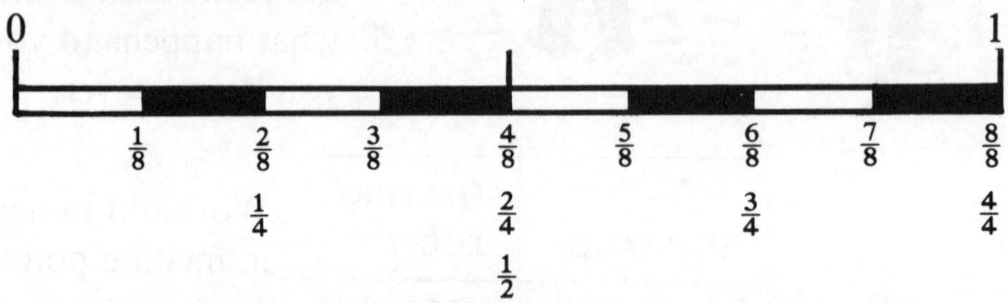

Use the eighths number line to write, in lowest terms:

1. $\frac{2}{8}$ **2.** $\frac{6}{8}$ **3.** $\frac{4}{8}$ **4.** $\frac{2}{4}$ **5.** $\frac{8}{8}$ **6.** $\frac{4}{4}$ **7.** $\frac{2}{2}$.

D

Complete the following:

1. $\frac{1}{2}=\frac{\square}{10}$ **2.** $\frac{1}{2}=\frac{\square}{8}$ **3.** $\frac{1}{2}=\frac{\square}{12}$ **4.** $\frac{1}{3}=\frac{\square}{12}$ **5.** $\frac{2}{3}=\frac{\square}{12}$

6. $\frac{1}{4}=\frac{\square}{8}$ **7.** $\frac{1}{4}=\frac{\square}{12}$ **8.** $\frac{3}{4}=\frac{\square}{12}$ **9.** $\frac{3}{4}=\frac{\square}{8}$ **10.** $\frac{5}{6}=\frac{\square}{12}$

11. $\frac{3}{5}=\frac{\square}{10}$ **12.** $\frac{2}{5}=\frac{\square}{10}$ **13.** $\frac{6}{6}=\square$ **14.** $\frac{4}{4}=\square$ **15.** $\frac{5}{5}=\square$

E Write in millimetres:

1. $\frac{6}{10}$ cm **2.** $\frac{3}{5}$ cm **3.** $\frac{4}{10}$ cm **4.** $\frac{2}{5}$ cm **5.** $\frac{1}{2}$ cm
6. $\frac{5}{10}$ cm **7.** $\frac{2}{10}$ cm **8.** $\frac{1}{5}$ cm **9.** $\frac{8}{10}$ cm **10.** $\frac{4}{5}$ cm

F Write in pence:

1. £ $\frac{1}{10}$ **2.** £ $\frac{2}{10}$ **3.** £ $\frac{1}{5}$ **4.** £ $\frac{4}{10}$ **5.** £ $\frac{2}{5}$
6. £ $\frac{5}{10}$ **7.** £ $\frac{1}{2}$ **8.** £ $\frac{6}{10}$ **9.** £ $\frac{3}{5}$ **10.** £ $\frac{4}{5}$

G Write in minutes:

1. $\frac{1}{10}$ of an hour **2.** $\frac{2}{10}$ of an hour **3.** $\frac{1}{5}$ of an hour
4. $\frac{5}{10}$ of an hour **5.** $\frac{1}{2}$ of an hour **6.** $\frac{6}{12}$ of an hour
7. $\frac{6}{10}$ of an hour **8.** $\frac{3}{5}$ of an hour **9.** $\frac{1}{4}$ of an hour
10. $\frac{1}{12}$ of an hour **11.** $\frac{2}{12}$ of an hour **12.** $\frac{1}{6}$ of an hour
13. $\frac{2}{6}$ of an hour **14.** $\frac{1}{3}$ of an hour **15.** $\frac{2}{3}$ of an hour

H $\frac{1}{5}$ of £1 is 20 p → $\frac{1}{5}$ of £3 is 20 p × 3 = 60 p.
Work out:

1. $\frac{2}{5}$ of £5 **2.** $\frac{1}{3}$ of £6 **3.** $\frac{2}{3}$ of £6 **4.** $\frac{1}{4}$ of £1
5. $\frac{1}{4}$ of £3 **6.** $\frac{3}{4}$ of £1 **7.** $\frac{3}{4}$ of £2 **8.** $\frac{3}{4}$ of £10
9. $\frac{3}{8}$ of £16 **10.** $\frac{3}{10}$ of £1 **11.** $\frac{3}{10}$ of £4 **12.** $\frac{1}{4}$ of £12.

I *Example:* What fraction of £10 is £2?

Write $\frac{\text{£2}}{\text{£10}} \rightarrow \frac{2}{10} \rightarrow \frac{1}{5}$.

What fraction of:

1. £20 is £4 **2.** £18 is £3 **3.** £18 is £6 **4.** £12 is £4
5. 50 p is 25 p **6.** 40 p is 8 p **7.** 60 p is 10 p **8.** 30 p is 15 p
9. 60 minutes is 5 minutes **10.** 1 hour is 5 minutes
11. 1 hour is 10 minutes **12.** 1 hour is 15 minutes
13. 24 hours is 6 hours **14.** 1 day is 6 hours
15. 1 day is 3 hours **16.** 1 day is 8 hours
17. 1 day is 4 hours **18.** 1 day is 12 hours
19. 1 day is 10 hours **20.** 1 day is 16 hours?

Improper fractions and mixed numbers

A

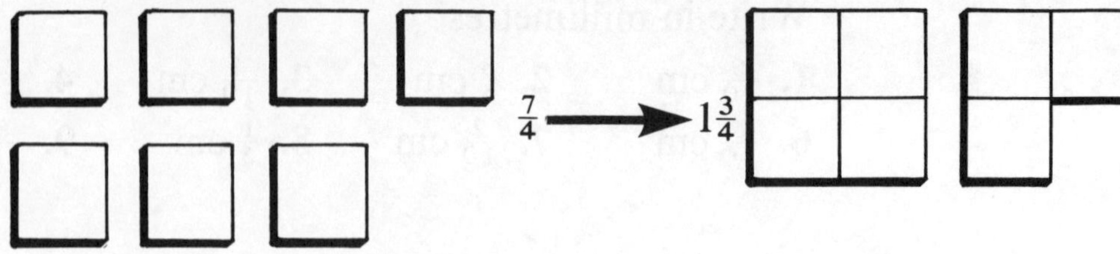

$\frac{7}{4}$ is an IMPROPER FRACTION $\quad$ $1\frac{3}{4}$ is a MIXED NUMBER

Express the shaded parts as IMPROPER FRACTIONS and as MIXED NUMBERS.

1.

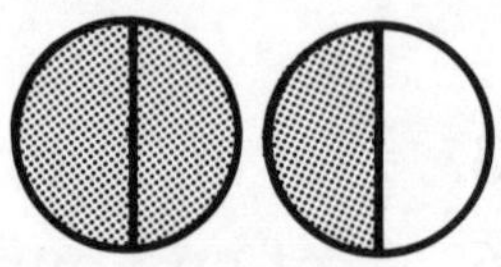

$\frac{3}{2} \rightarrow \square$

2.

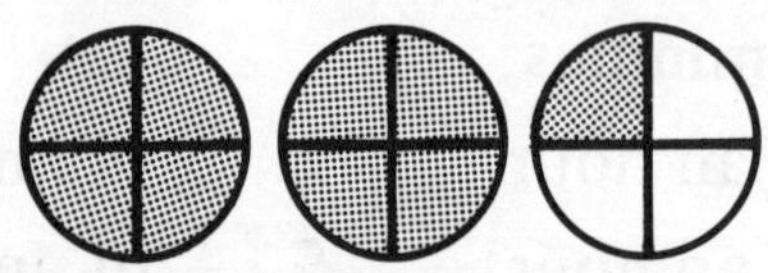

3.

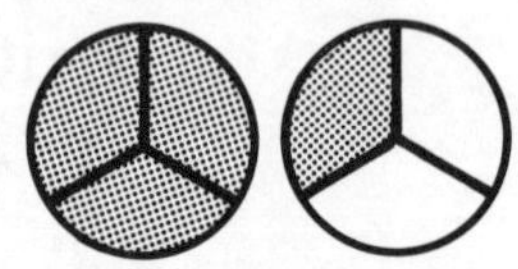

4.

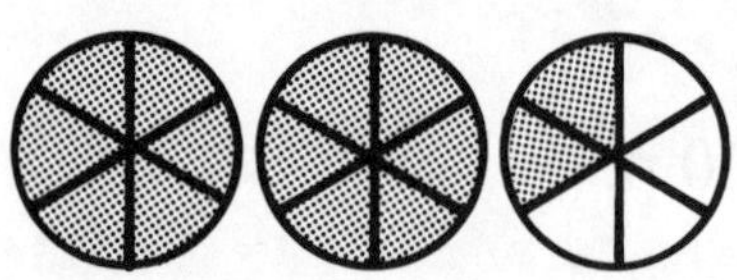

5.

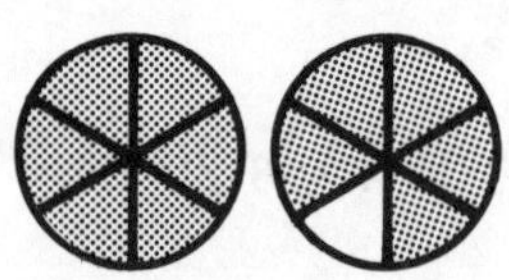

6.

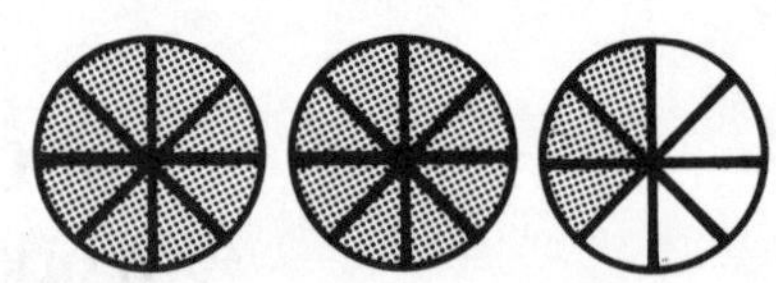

B Copy and complete the table.

centimetres	1	$1\frac{1}{2}$	2			$3\frac{1}{2}$			5	
$\frac{1}{2}$ centimetres	2			5	6			9		11
millimetres		15					40			

C Write as whole numbers (the first is done):

1. $\frac{8}{2} = \mathbf{4}$ $\quad$ **2.** $\frac{6}{3} = \square$ $\quad$ **3.** $\frac{12}{4} = \square$ $\quad$ **4.** $\frac{20}{5} = \square$ $\quad$ **5.** $\frac{20}{4} = \square$

6. $\frac{24}{8} = \square$ $\quad$ **7.** $\frac{15}{3} = \square$ $\quad$ **8.** $\frac{18}{6} = \square$ $\quad$ **9.** $\frac{16}{2} = \square$ $\quad$ **10.** $\frac{24}{6} = \square$

11. $\frac{24}{2} = \square$ $\quad$ **12.** $\frac{12}{6} = \square$ $\quad$ **13.** $\frac{15}{5} = \square$ $\quad$ **14.** $\frac{9}{3} = \square$ $\quad$ **15.** $\frac{21}{3} = \square$

D

Examples: **1 can be written $\frac{2}{2}, \frac{3}{3}, \frac{4}{4}, \frac{5}{5}, \cdots$**
2 can be written $\frac{4}{2}, \frac{6}{3}, \frac{8}{4}, \frac{10}{5}, \cdots$

Write each of the following as halves, thirds, quarters, fifths, sixths and so on.

1. 3 **2.** 4 **3.** 5 **4.** 6 **5.** 7 **6.** 8

E

Example: Change $3\frac{1}{2}$ to an IMPROPER FRACTION.
$3\frac{1}{2} \rightarrow 3 + \frac{1}{2} \rightarrow \frac{6}{2} + \frac{1}{2} \rightarrow \frac{7}{2}$.

Change to IMPROPER FRACTIONS:

1. $1\frac{1}{2}$ **2.** $3\frac{1}{2}$ **3.** $2\frac{1}{4}$ **4.** $2\frac{3}{4}$ **5.** $2\frac{1}{3}$
6. $3\frac{3}{8}$ **7.** $2\frac{5}{8}$ **8.** $6\frac{1}{4}$ **9.** $5\frac{2}{3}$ **10.** $4\frac{2}{3}$
11. $1\frac{4}{5}$ **12.** $3\frac{1}{7}$ **13.** $3\frac{7}{8}$ **14.** $2\frac{5}{6}$ **15.** $4\frac{1}{6}$
16. $1\frac{2}{5}$ **17.** $3\frac{1}{5}$ **18.** $6\frac{1}{10}$ **19.** $4\frac{9}{10}$ **20.** $2\frac{4}{5}$

F

Example: Change $\frac{5}{3}$ to a MIXED NUMBER.
$\frac{5}{3}$ means $5 \div 3 \rightarrow 1$ rem 2 or $1\frac{2}{3}$ (2 divided by 3).

Change to MIXED NUMBERS in their lowest terms.

1. $\frac{5}{4}$ **2.** $\frac{7}{4}$ **3.** $\frac{10}{3}$ **4.** $\frac{9}{4}$ **5.** $\frac{11}{5}$
6. $\frac{18}{4}$ **7.** $\frac{18}{10}$ **8.** $\frac{20}{6}$ **9.** $\frac{26}{8}$ **10.** $\frac{30}{8}$
11. $\frac{22}{7}$ **12.** $\frac{14}{4}$ **13.** $\frac{12}{10}$ **14.** $\frac{11}{3}$ **15.** $\frac{26}{5}$
16. $\frac{21}{5}$ **17.** $\frac{33}{7}$ **18.** $\frac{22}{4}$ **19.** $\frac{34}{8}$ **20.** $\frac{23}{3}$

Adding fractions

A Copy these number lines and complete the adding.

1.

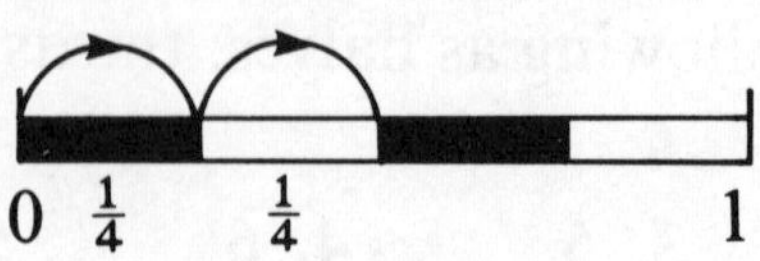

$$\frac{1}{4}+\frac{1}{4}=\frac{\square}{4}=\frac{1}{\square}$$

2.

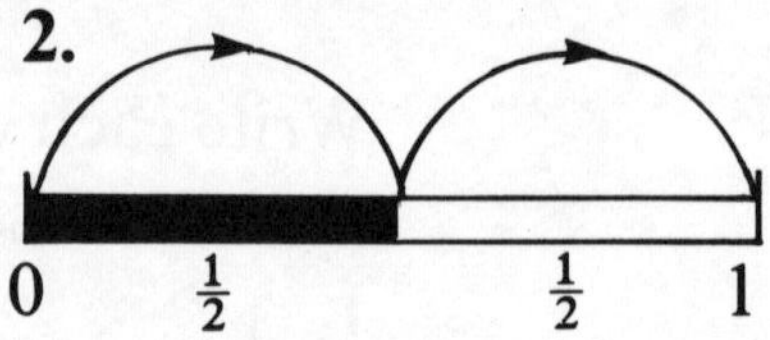

$$\frac{1}{2}+\frac{1}{2}=\frac{\square}{2}=\square$$

3.

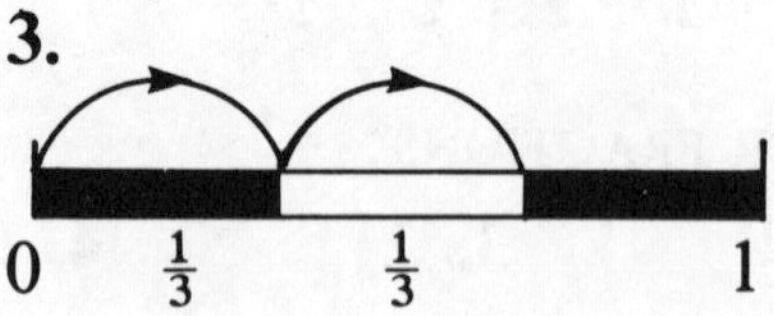

$$\frac{1}{3}+\frac{1}{3}=\frac{\square}{3}$$

4.

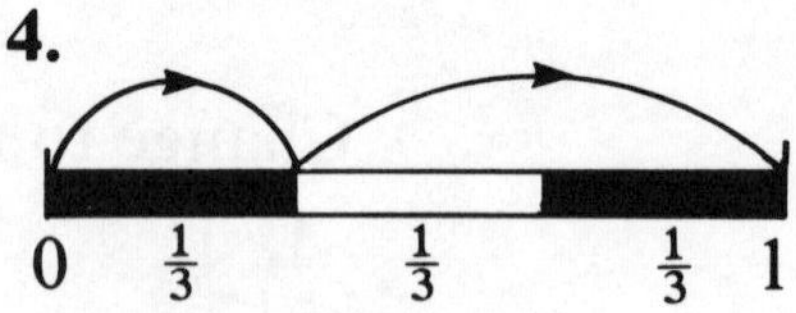

$$\frac{1}{3}+\frac{2}{3}=\frac{\square}{3}=\square$$

5.

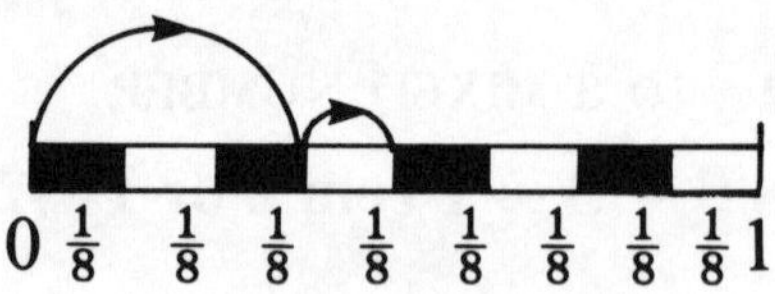

$$\frac{3}{8}+\frac{1}{8}=\frac{\square}{8}=\frac{1}{\square}$$

6.

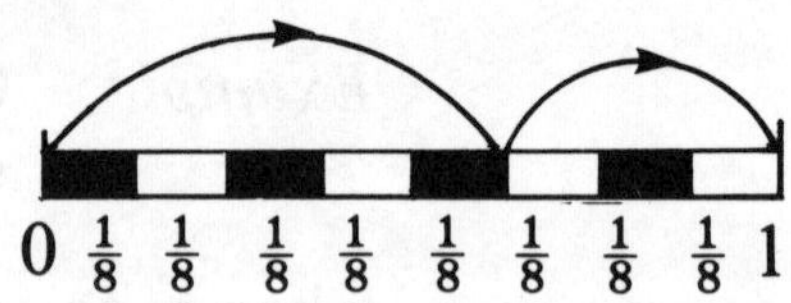

$$\frac{5}{8}+\frac{3}{8}=\frac{\square}{8}=\square$$

B Add:

1. $\frac{1}{3}+\frac{1}{3}=\frac{\square}{3}$

2. $\frac{1}{6}+\frac{1}{6}=\frac{\square}{6}=\frac{1}{\square}$

3. $\frac{1}{4}+\frac{1}{4}=\frac{\square}{4}=\frac{1}{\square}$

4. $\frac{2}{5}+\frac{2}{5}=\frac{\square}{5}$

5. $\frac{1}{8}+\frac{1}{8}=\frac{\square}{8}=\frac{1}{\square}$

6. $\frac{1}{10}+\frac{1}{10}=\frac{\square}{10}=\frac{1}{\square}$

7. $\frac{1}{7}+\frac{1}{7}=\frac{\square}{7}$

8. $\frac{5}{8}+\frac{1}{8}=\frac{\square}{8}=\frac{3}{\square}$

9. $\frac{1}{6}+\frac{5}{6}=\frac{\square}{6}=\square$

10. $\frac{2}{5}+\frac{1}{5}=\frac{\square}{5}$

11. $\frac{3}{10}+\frac{1}{10}=\frac{\square}{10}=\frac{2}{\square}$

12. $\frac{1}{8}+\frac{3}{8}=\frac{\square}{8}=\frac{1}{\square}$

13. $\frac{1}{9}+\frac{1}{9}=\frac{\square}{9}$

14. $\frac{1}{10}+\frac{7}{10}=\frac{\square}{10}=\frac{4}{\square}$

15. $\frac{1}{12}+\frac{7}{12}=\frac{\square}{12}=\frac{2}{\square}$

16. $\frac{1}{5}+\frac{3}{5}=\frac{\square}{5}$

17. $\frac{3}{8}+\frac{3}{8}=\frac{\square}{8}=\frac{3}{\square}$

18. $\frac{2}{9}+\frac{1}{9}=\frac{\square}{9}=\frac{1}{\square}$

C A journey takes $\frac{1}{4}$ hour there and $\frac{1}{4}$ hour back.

1. How long is this altogether?
2. How long will 4 single journeys (one way) take?
3. How long will 5 single journeys (one way) take?
4. How long will 5 *return* journeys take?

D Complete these mappings.

1.
- $\frac{1}{2} \rightarrow \frac{\square}{4}$
- $\frac{1}{4} \rightarrow \frac{\square}{8}$
- $\frac{3}{4} \rightarrow \frac{\square}{8}$
- $\frac{1}{2} \rightarrow \frac{\square}{10}$
- $\frac{1}{4} \rightarrow \frac{\square}{12}$
- $\frac{1}{2} \rightarrow \frac{\square}{6}$

2.
- $\frac{1}{3} \rightarrow \frac{\square}{6}$
- $\frac{1}{6} \rightarrow \frac{\square}{12}$
- $\frac{1}{2} \rightarrow \frac{\square}{12}$
- $\frac{1}{3} \rightarrow \frac{\square}{12}$
- $\frac{2}{3} \rightarrow \frac{\square}{12}$
- $\frac{2}{3} \rightarrow \frac{\square}{6}$

3.
- $\frac{3}{4} \rightarrow \frac{\square}{12}$
- $\frac{5}{6} \rightarrow \frac{\square}{12}$
- $\frac{1}{5} \rightarrow \frac{\square}{10}$
- $\frac{3}{5} \rightarrow \frac{\square}{10}$
- $\frac{4}{5} \rightarrow \frac{\square}{10}$
- $\frac{2}{3} \rightarrow \frac{\square}{9}$

E *Examples:* **1.** $\frac{1}{2}+\frac{1}{4}\rightarrow\frac{2}{4}+\frac{1}{4}\rightarrow\frac{3}{4}$ **2.** $\frac{1}{3}+\frac{1}{6}\rightarrow\frac{2}{6}+\frac{1}{6}\rightarrow\frac{3}{6}\rightarrow\frac{1}{2}$.

Add, and give the answers in their lowest terms.

1. $\frac{1}{4}+\frac{1}{8}$ **2.** $\frac{1}{2}+\frac{1}{8}$ **3.** $\frac{1}{5}+\frac{1}{10}$ **4.** $\frac{1}{2}+\frac{1}{6}$ **5.** $\frac{1}{4}+\frac{1}{12}$

6. $\frac{1}{2}+\frac{1}{10}$ **7.** $\frac{1}{3}+\frac{1}{12}$ **8.** $\frac{1}{4}+\frac{1}{16}$ **9.** $\frac{1}{16}+\frac{1}{2}$ **10.** $\frac{1}{6}+\frac{1}{3}$

11. $\frac{1}{9}+\frac{1}{3}$ **12.** $\frac{1}{20}+\frac{1}{4}$ **13.** $\frac{1}{4}+\frac{3}{8}$ **14.** $\frac{1}{5}+\frac{3}{10}$ **15.** $\frac{1}{2}+\frac{3}{8}$

16. $\frac{1}{12}+\frac{5}{6}$ **17.** $\frac{1}{10}+\frac{2}{5}$ **18.** $\frac{3}{10}+\frac{1}{2}$ **19.** $\frac{5}{8}+\frac{1}{4}$ **20.** $\frac{1}{3}+\frac{7}{12}$

21. $\frac{2}{3}+\frac{1}{6}$ **22.** $\frac{1}{4}+\frac{3}{16}$ **23.** $\frac{3}{4}+\frac{1}{8}$ **24.** $\frac{1}{12}+\frac{2}{3}$ **25.** $\frac{1}{8}+\frac{1}{2}$

F Find the length in hours of these journeys.
Complete the table.

1ST STAGE — TIME IN HOURS 2ND STAGE — TIME IN HOURS	$\frac{1}{4}$ $\frac{1}{2}$	$\frac{1}{2}$ $\frac{1}{10}$	$\frac{1}{3}$ $\frac{1}{6}$	$\frac{1}{4}$ $\frac{1}{12}$	$\frac{1}{5}$ $\frac{3}{10}$	$\frac{3}{10}$ $\frac{1}{2}$	$\frac{1}{4}$ $\frac{3}{20}$	$\frac{1}{3}$ $\frac{5}{12}$
TOTAL TIME IN HOURS	$\frac{3}{4}$							
TOTAL TIME IN MINUTES	**45**							

Subtracting fractions

A Copy the number lines and then complete the taking away.

1.

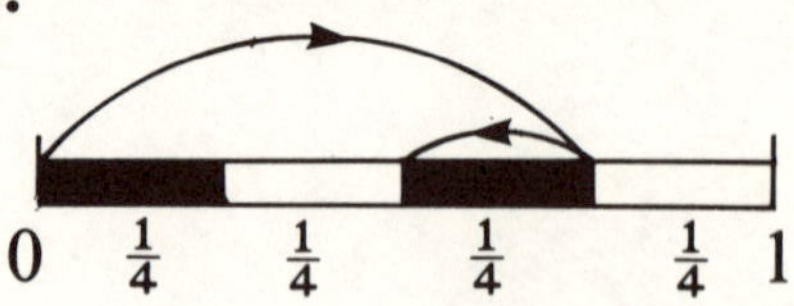

$\frac{3}{4} - \frac{1}{4} = \frac{\square}{4} = \frac{1}{\square}$

2.

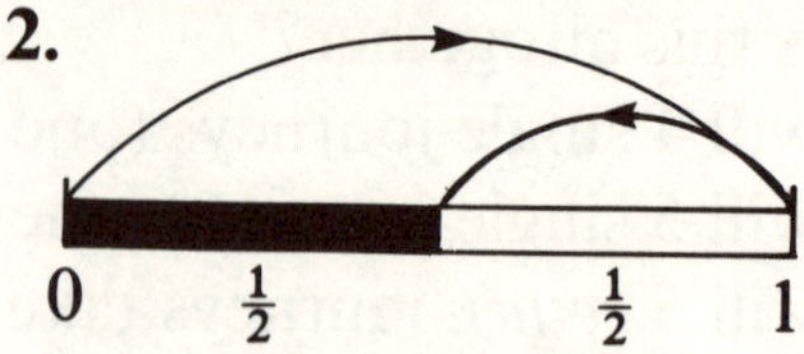

$\square \rightarrow \frac{2}{2} - \frac{1}{2} = \frac{\square}{2}$

3.

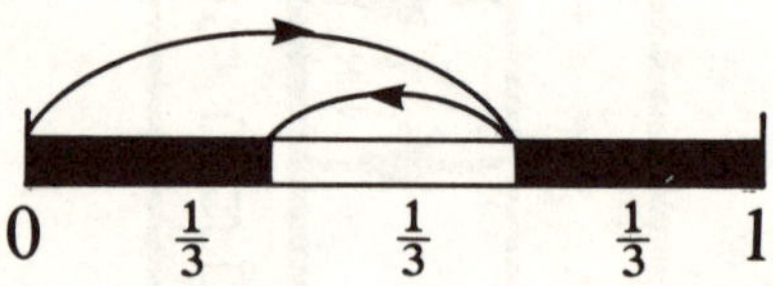

$\frac{2}{3} - \frac{1}{3} = \frac{\square}{3}$

4.

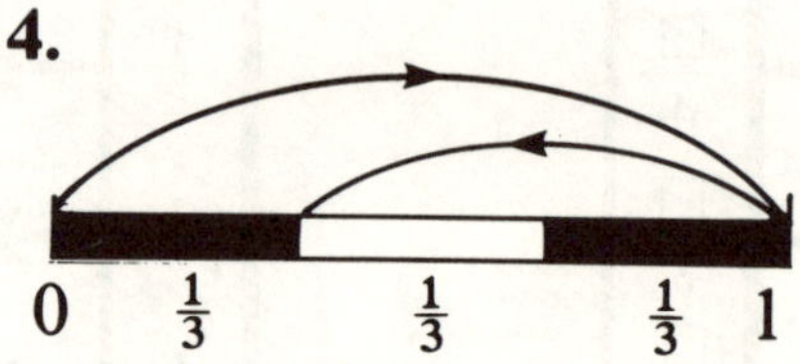

$\square \rightarrow \frac{3}{3} - \frac{2}{3} = \frac{\square}{3}$

5.

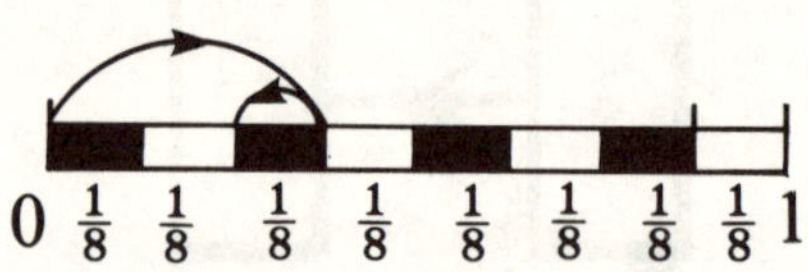

$\frac{3}{8} - \frac{1}{8} = \frac{\square}{8} = \frac{1}{\square}$

6.

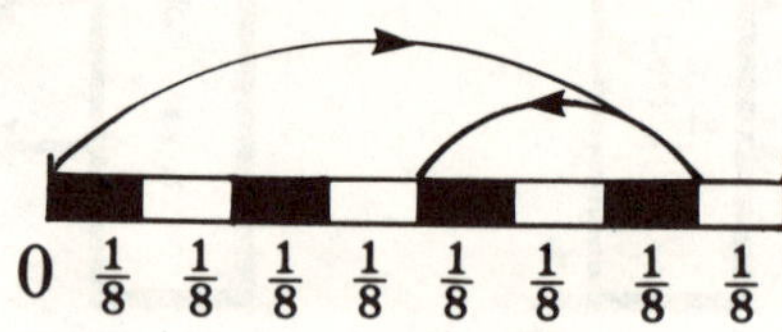

$\frac{7}{8} - \frac{3}{8} = \frac{\square}{8} = \frac{1}{\square}$

B Work out the following. Write the answers in their lowest terms.

1. $\frac{5}{6} - \frac{1}{6} = \frac{\square}{6} = \frac{\square}{3}$ **2.** $\frac{7}{8} - \frac{1}{8} = \frac{\square}{8} = \frac{\square}{\square}$ **3.** $\frac{7}{12} - \frac{5}{12} = \frac{\square}{12} = \frac{\square}{\square}$

4. $\frac{4}{5} - \frac{1}{5} = \frac{\square}{5}$ **5.** $\frac{9}{10} - \frac{3}{10} = \frac{\square}{10} = \frac{\square}{\square}$ **6.** $\frac{7}{8} - \frac{5}{8} = \frac{\square}{8} = \frac{\square}{\square}$

7. $\frac{3}{5} - \frac{1}{5}$ **8.** $\frac{7}{10} - \frac{3}{10}$ **9.** $\frac{4}{5} - \frac{3}{5}$ **10.** $\frac{11}{12} - \frac{1}{12}$ **11.** $\frac{9}{10} - \frac{3}{10}$

12. $\frac{5}{8} - \frac{3}{8}$ **13.** $\frac{9}{10} - \frac{1}{10}$ **14.** $\frac{7}{12} - \frac{5}{12}$ **15.** $\frac{7}{10} - \frac{1}{10}$ **16.** $\frac{5}{12} - \frac{1}{12}$

C Work out:

1. $1 - \frac{1}{2}$ **2.** $1 - \frac{1}{3}$ **3.** $1 - \frac{1}{4}$ **4.** $1 - \frac{1}{5}$ **5.** $1 - \frac{1}{8}$

6. $1 - \frac{3}{8}$ **7.** $1 - \frac{5}{6}$ **8.** $1 - \frac{7}{12}$ **9.** $1 - \frac{3}{10}$ **10.** $1 - \frac{7}{8}$

11. $1 - \frac{2}{3}$ **12.** $1 - \frac{3}{4}$ **13.** $1 - \frac{1}{6}$ **14.** $1 - \frac{1}{12}$ **15.** $1 - \frac{7}{10}$

D Complete the mappings of EQUIVALENT FRACTIONS.

1.

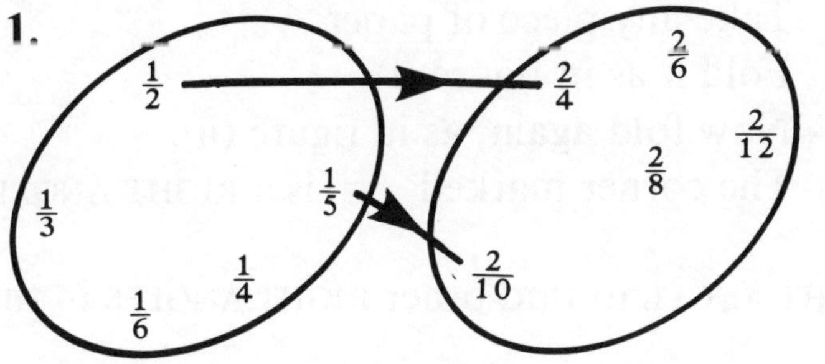

2.

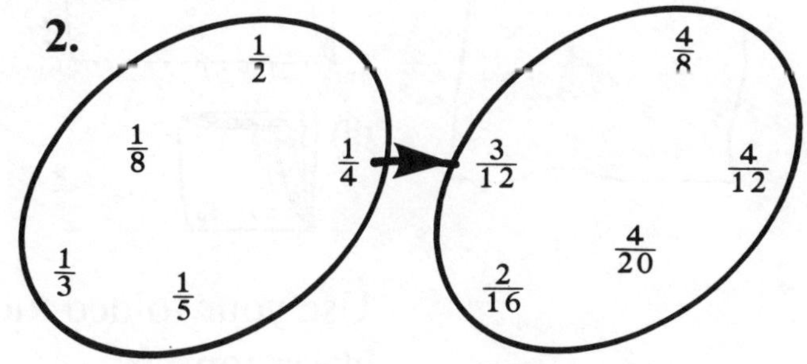

3.

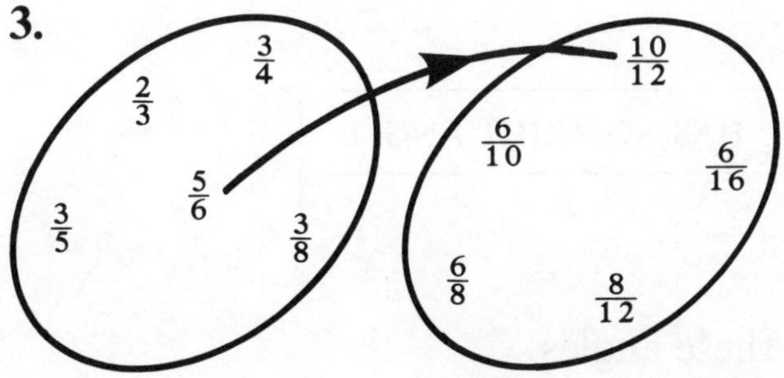

4.

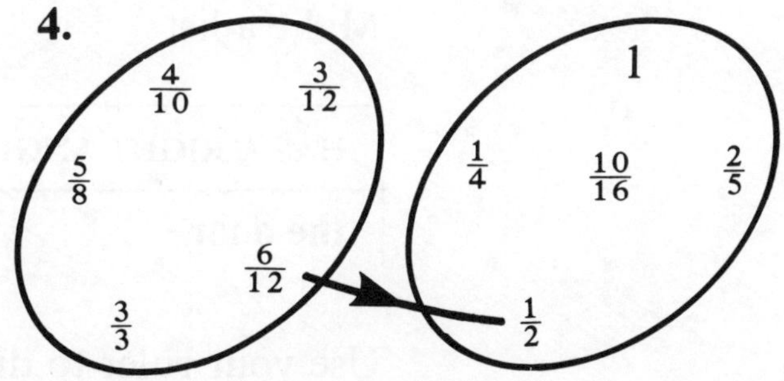

E Work out. (The first is done for you.)

1. $\frac{1}{2} - \frac{1}{4} \rightarrow \mathbf{\frac{2}{4} - \frac{1}{4} = \frac{1}{4}}$

2. $\frac{1}{2} - \frac{1}{6}$ 3. $\frac{1}{2} - \frac{1}{10}$ 4. $\frac{1}{2} - \frac{1}{8}$ 5. $\frac{1}{2} - \frac{1}{12}$

6. $\frac{1}{4} - \frac{1}{8}$ 7. $\frac{1}{3} - \frac{1}{12}$ 8. $\frac{1}{5} - \frac{1}{10}$ 9. $\frac{1}{3} - \frac{1}{6}$

10. $\frac{1}{4} - \frac{1}{12}$ 11. $\frac{3}{4} - \frac{1}{2}$ 12. $\frac{5}{8} - \frac{1}{4}$ 13. $\frac{5}{8} - \frac{1}{16}$

14. $\frac{7}{10} - \frac{1}{2}$ 15. $\frac{5}{6} - \frac{1}{3}$ 16. $\frac{3}{10} - \frac{1}{5}$ 17. $\frac{5}{6} - \frac{1}{2}$

18. $\frac{5}{8} - \frac{1}{2}$ 19. $\frac{9}{10} - \frac{3}{5}$ 20. $\frac{7}{8} - \frac{3}{4}$ 21. $\frac{5}{6} - \frac{2}{3}$

22. $\frac{3}{4} - \frac{3}{8}$ 23. $\frac{7}{10} - \frac{3}{5}$ 24. $\frac{3}{5} - \frac{3}{10}$ 25. $\frac{3}{4} - \frac{5}{12}$

F Brian lived $\frac{3}{4}$ km from school. How far did he still have to go to school when he had covered $\frac{3}{8}$ km?
How far had he covered when he had $\frac{1}{8}$ km still to go?

Perpendicular and parallel lines

A

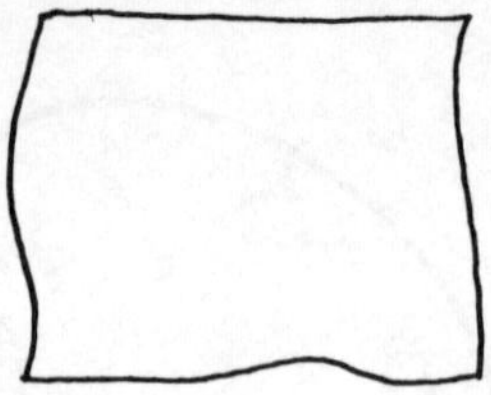

(i) 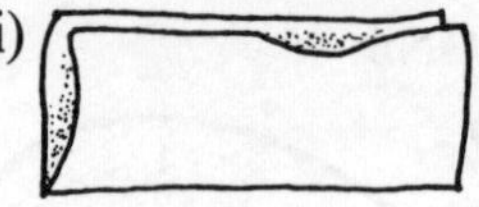

(ii) 

You can fold a RIGHT ANGLE.
Take any piece of paper.
Fold it as in figure (i).
Now fold again, as in figure (ii).
The corner marked ⊡ is a RIGHT ANGLE.

Use your folded RIGHT ANGLE to find other RIGHT ANGLES in the classroom.
Make a list:

HAS A RIGHT ANGLE	HAS NO RIGHT ANGLE
the door	

B

Use your ruler to draw these angles.

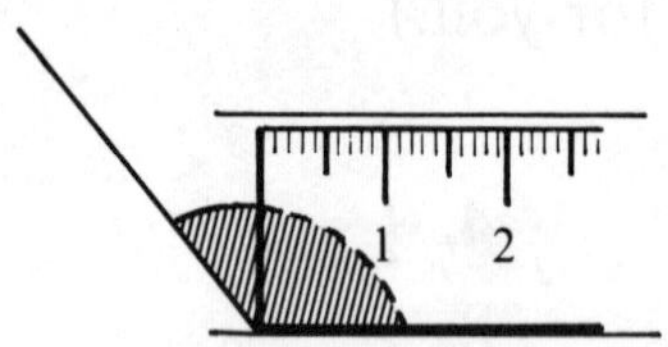

obtuse angle

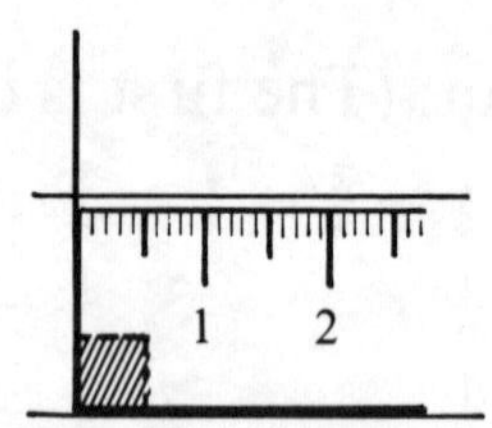

right angle

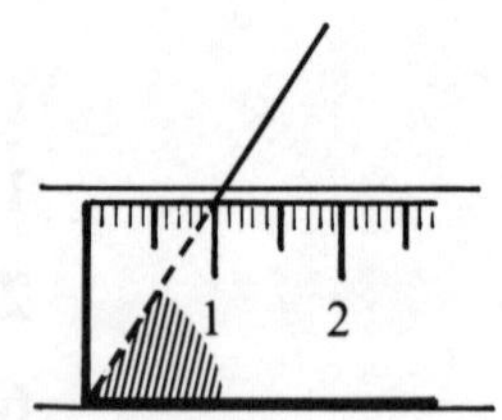

acute angle

Any two lines which make a RIGHT ANGLE are PERPENDICULAR to each other.
The middle pair of lines are PERPENDICULAR.

Use your ruler to find which of these pairs of lines are PERPENDICULAR.

1.

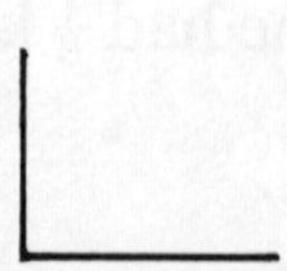

2.

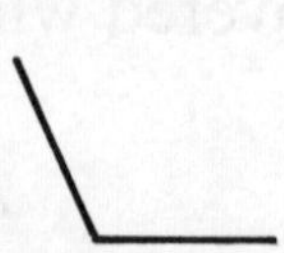

3.

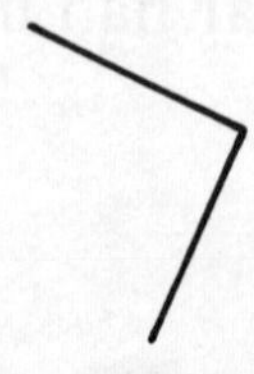

4.

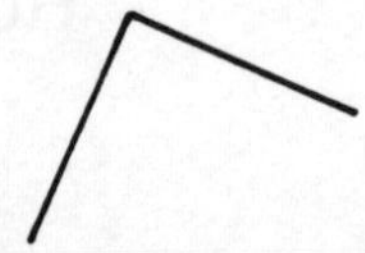

5.

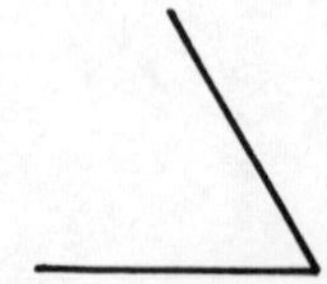

6.

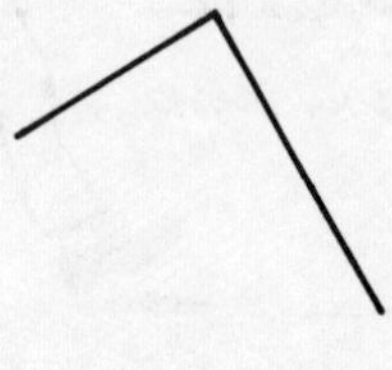

7.

8.

9.

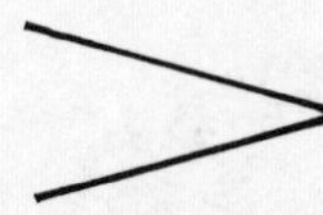

10.

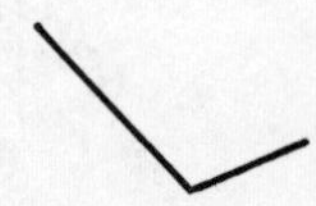

C

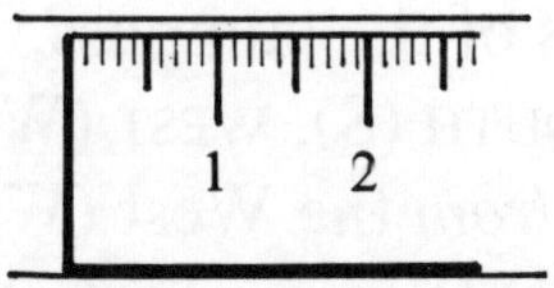

PARALLEL lines stay the same distance apart.
Drawing either side of a ruler gives a pair of PARALLEL lines.
The angle between PARALLEL lines is 0°.

Here is a set of kitchen cupboards.
Copy them.

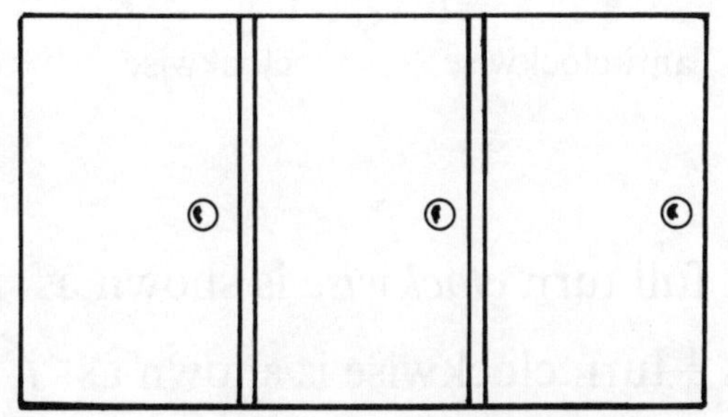

1. Mark with an arrow —→— one set of lines that are PARALLEL.
2. Mark with an arrow —→→— another set of lines that are PARALLEL.
3. Mark in the corners of your drawing the pairs of lines that are PERPENDICULAR.

D

Look at these shapes. Then copy and complete the table.

1.
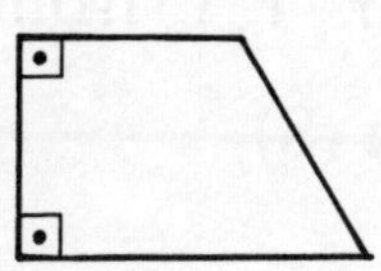

2.
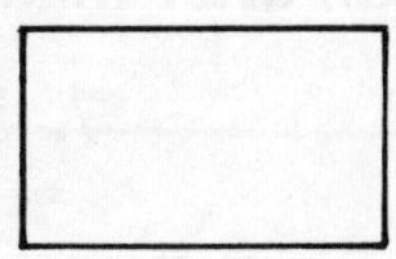

3.
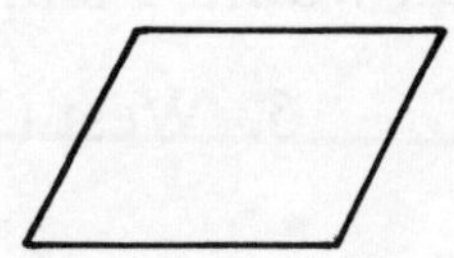

4.
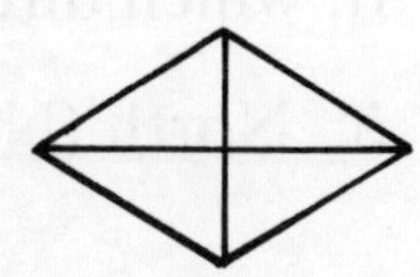

5.
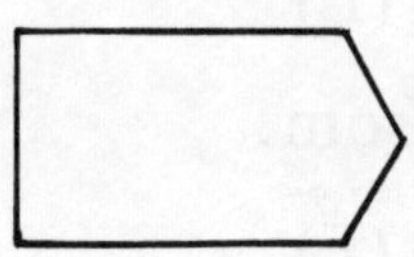

6.

7.
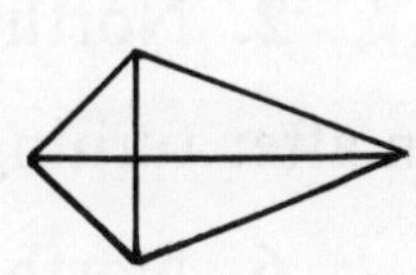

8.
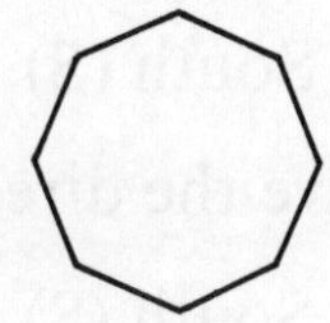

SHAPE	NUMBER OF PAIRS OF PERPENDICULAR LINES	NUMBER OF PAIRS OF PARALLEL LINES
1	**2**	**1**
2		
3		
4		
5		
6		
7		
8		

Directions and turns

A

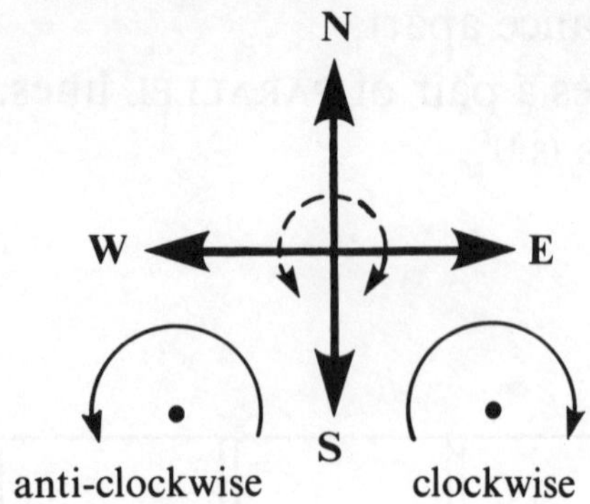

There are 4 main points of the compass, NORTH (N), EAST (E), SOUTH (S), WEST (W).
A westerly wind blows *from* the West (W).
It blows *towards* the East (E). → E

Towards which direction do these winds blow:

1. northerly **2.** easterly **3.** southerly?

B A full turn *clockwise* is shown as (1 ↷).

A $\frac{1}{4}$ turn clockwise is shown as ($\frac{1}{4}$ ↷). A $\frac{1}{4}$ turn anti-clockwise is shown as

In which direction would a ship sail after making ($\frac{1}{4}$ ↷) from:

1. North (N) **2.** West (W) **3.** South (S) **4.** East (E)?

In which direction would a ship sail after making (↶ $\frac{1}{4}$) from:

5. North (N) **6.** West (W) **7.** South (S) **8.** East (E)?

C Give the direction after turning ($\frac{1}{2}$ ↷) starting from:

1. South (S) **2.** North (N) **3.** East (E) **4.** West (W).

Give the direction after turning (↶ $\frac{1}{2}$) starting from:

5. South (S) **6.** North (N) **7.** East (E) **8.** West (W).

9. What do you notice about these two half turns?

D Record the result of making this turn ($\frac{3}{4}$ ↷) from:

1. East (E) **2.** North (N)

3. South (S) **4.** West (W).

5. Starting at the same directions, make a (↶ $\frac{3}{4}$) turn from each.

Record your results.

E Starting from North (N), make the following combined turns. Some are anti-clockwise. Record each result.

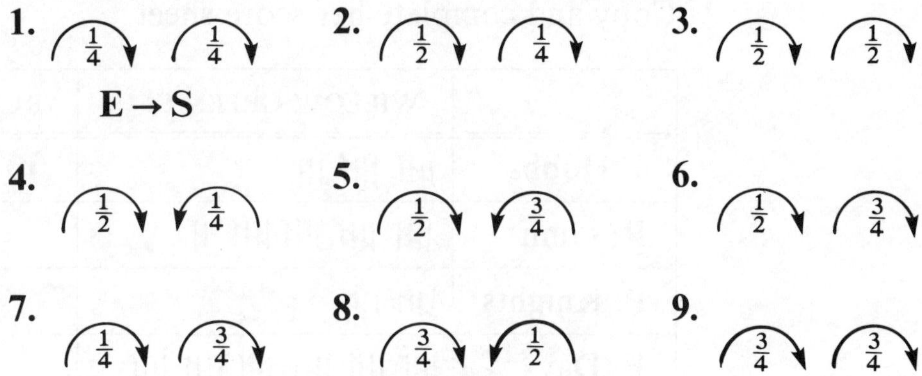

10. For each question, which single turn would give the same result?
This is number 1: 

F A ship which makes a $\frac{1}{4}$ turn from North (N) turns through 1 RIGHT ANGLE.

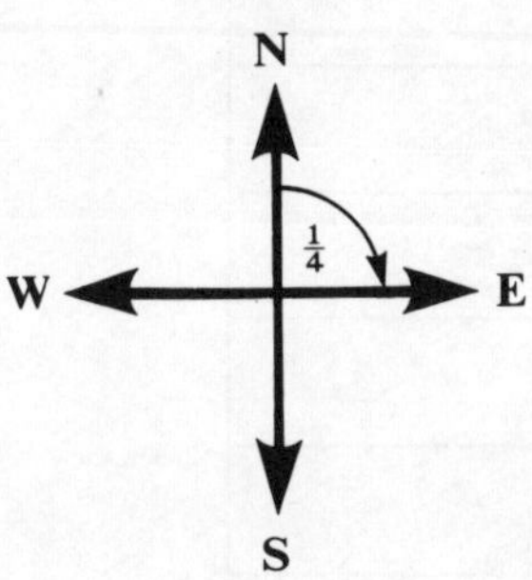

In which direction will the ship sail after turning clockwise:

1. 2 right angles from N
2. 1 right angle from W
3. 2 right angles from E
4. 1 right angle from S
5. 3 right angles from E
6. 3 right angles from W
7. 4 right angles from N?

G When a ship has turned 4 RIGHT ANGLES it has turned a full circle.
Complete these sentences.

1. 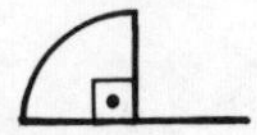 1 RIGHT ANGLE is $\frac{1}{\square}$ of a full circle.
2. 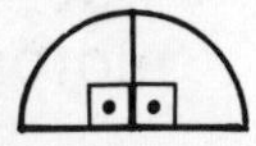 2 RIGHT ANGLES make $\frac{1}{\square}$ of a full circle.
3. 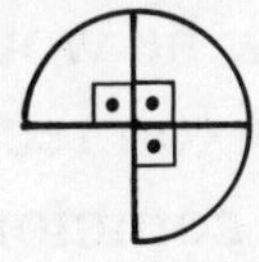 3 RIGHT ANGLES make $\frac{\square}{\square}$ of a full circle.

Tallying and pictograms

A Mary Hobbs kept score for her village cricket team.
She made a tally mark, , for one run, and for 5 runs.
Copy and complete her score sheet.

	WILLOW GREEN C. C.	RUNS SCORED
J. Hobbs	𝍸 𝍸 |||	**13**
R. Ginn	𝍸 𝍸 𝍸 𝍸 ||	
P. Knights	𝍸 ||||	
F. Day	𝍸 𝍸 𝍸 𝍸 𝍸 ||||	
D. Pound		
S. Pence	𝍸 ||	
J. Hunt	𝍸 𝍸		
C. Ball	𝍸 𝍸 𝍸	
A. Cross	𝍸 𝍸 𝍸		
F. Patch	𝍸 |||	
D. Amuah	𝍸 ||	
EXTRAS	𝍸 ||	
	TOTAL SCORE	

Who made the highest score?

B

P.C. Dixon made a survey of the vehicles passing through his village.
His results were as follows:

private cars	27
vans	21
lorries	12
motor cycles	15
bicycles	29

1. Make a tally chart to show how he would record these totals.
2. How many vehicles did P.C. Dixon record altogether?
3. What type of vehicle was most common?

C Work out the totals represented by these pictograms.

1.  Each umbrella represents 7 rainy days → **4 × 7 = 28 rainy days.**
2.  Each suitcase represents 30 aircraft passengers.
3.  Each pram represents 10 new-born babies.
4.  Each tree represents 150 newly planted conifers.
5.  Each mouse represents 25 mice caught.
6.  Each house represents 250 new houses built.
7.  Each car represents 12 cars sold.

D Draw pictograms to represent:

1. 36 sheep —each picture to represent 6 sheep.
2. 24 aeroplanes—each picture to represent 8 aeroplanes.
3. 54 sunny days—each picture to represent 9 sunny days.
4. 450 cabbages —each picture to represent 50 cabbages.
5. 1000 soldiers —each picture to represent 100 soldiers.

E How many does each of these pictures represent?

1.  150 cups of tea sold altogether.
2.  240 kg of fish caught.
3.  1500 litres of milk collected.
4.  175 chickens reared.
5.  72 teachers employed.

F Each football represents 5 goals scored by these teams.
Work out the total goals scored by each team.

Everton  **→ 5 × 10 = 50**
Derby 
Coventry 
Bolton 
Arsenal 

Which team scored fewest goals? How many fewer than the next?

Decimal fractions – tenths and hundredths

A

The dials of the electricity meter read $7642\frac{1}{10}$ units.
The meter-reader writes on his sheet **7642·1 units.**
What should he write for these readings?

1. $5252\frac{3}{10}$ units　**2.** $6724\frac{5}{10}$ units　**3.** $9263\frac{7}{10}$ units
4. $4776\frac{2}{10}$ units　**5.** $7452\frac{6}{10}$ units　**6.** $8715\frac{1}{10}$ units
7. $9463\frac{4}{10}$ units　**8.** $5327\frac{0}{10}$ units　**9.** $0029\frac{9}{10}$ units

B What is the shaded part as a decimal fraction of the shape?

1.
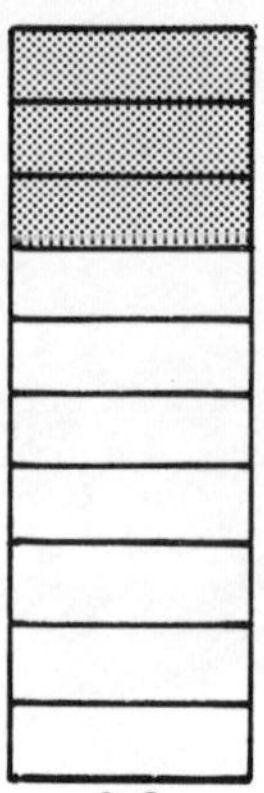
0·3

2.
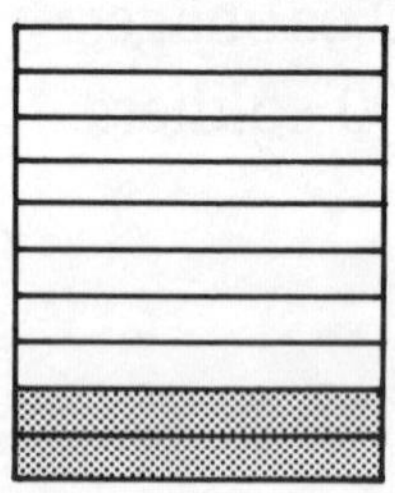

3.

4.
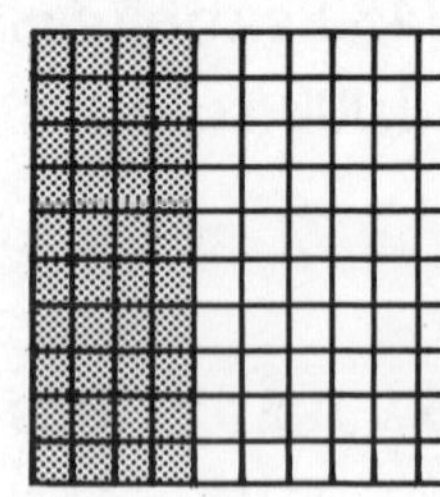

5.
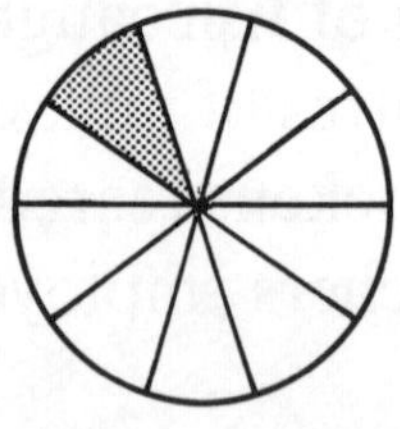

6.
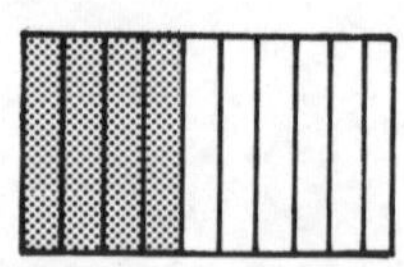

7. Now record as a decimal fraction the unshaded part of each shape.

C Write as mixed numbers, like this: **6·3 → $6\frac{3}{10}$.**

1. 4·5　**2.** 5·7　**3.** 8·9　**4.** 7·2　**5.** 10·4　**6.** 10·8

D Write as decimal fractions, like this: **$\frac{4}{10}$ → 0·4.**

1. $\frac{7}{10}$　**2.** $\frac{3}{10}$　**3.** $\frac{5}{10}$　**4.** $1\frac{7}{10}$　**5.** $2\frac{9}{10}$　**6.** $3\frac{3}{10}$

E

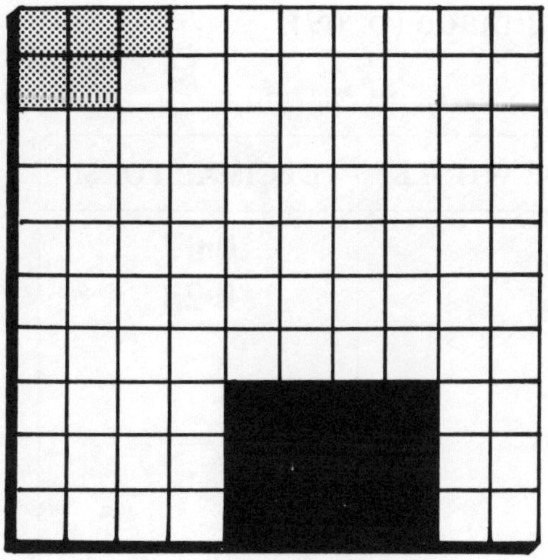

In the large square, 5 small squares out of 100 are shaded → $\frac{5}{100}$ or **0·05**.

12 squares out of the 100 are blacked in → $\frac{12}{100}$ or **0·12**.

On squared paper draw large squares with 100 small squares.

Shade:

1. 25 small squares **2.** 50 small squares **3.** 75 small squares.

Write the areas you have shaded as: (a) common fractions
(b) decimal fractions

> It is important to remember these: $\frac{25}{100} = \frac{1}{4} = 0{\cdot}25$; $\frac{50}{100} = \frac{1}{2} = 0{\cdot}5$; $\frac{75}{100} = \frac{3}{4} = 0{\cdot}75$

F

Copy and complete these mappings.

1.

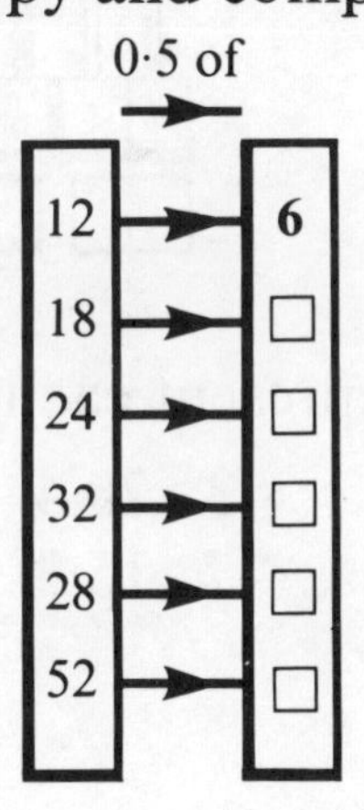

2.

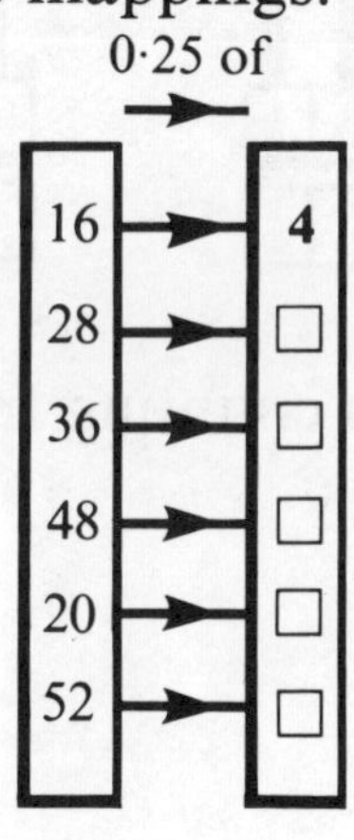

3.

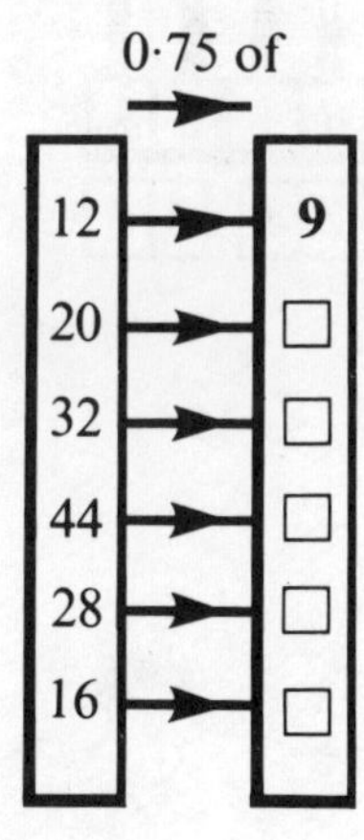

G Fill in the boxes.

1. $\frac{1}{2} = \frac{5}{10} = 0{\cdot}\boxed{5}$ **2.** $\frac{1}{20} = \frac{5}{100} = 0{\cdot}\boxed{0}\boxed{5}$ **3.** $\frac{1}{5} = \frac{2}{10} = 0{\cdot}\square$

4. $\frac{1}{50} = \frac{2}{100} = 0{\cdot}\square\square$ **5.** $\frac{3}{10} = \square{\cdot}\square$ **6.** $\frac{3}{100} = 0{\cdot}\square\square$

7. $\frac{2}{5} = \frac{\square}{10} = 0{\cdot}\square$ **8.** $\frac{2}{50} = \frac{\square}{100} = 0{\cdot}\square\square$ **9.** $\frac{1}{25} = \frac{\square}{100} = 0{\cdot}\square\square$

More about decimals

A Look at the hundred squares you made (p. 69).
Copy and complete the table.

SMALL SQUARES	FRACTION OF WHOLE	DECIMAL FORM
7	$\frac{7}{100}$	**0·07**
28	$\frac{28}{100}$	**0·28**
30		
69		
4		
6		
70		
25		
75		
90		

B Fill in the boxes below each counting frame.

1.

2.

3.
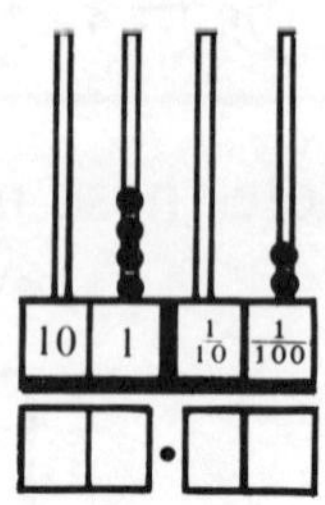

4.
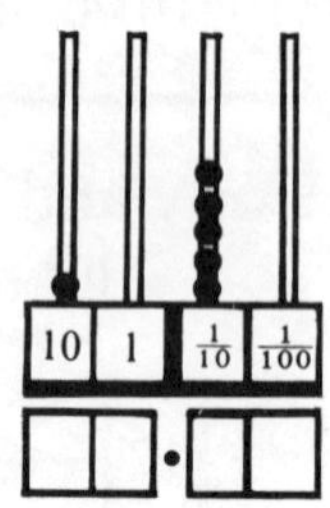

5.
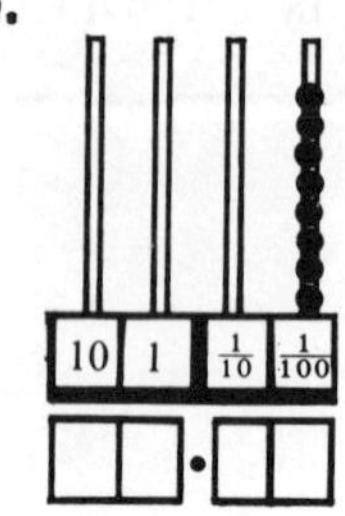

C Draw a counting frame to represent each of these.

1. 26·85 **2.** 7·16 **3.** 20·4 **4.** 0·13
5. 9·02 **6.** 0·07 **7.** 15·3 **8.** 10·04

D

Who jumped highest or farthest, or ran fastest on sports day?

1. High jump — Alan 1·30 m, Bob 1·25 m
2. Long jump — Karen 3·09 m, Liza 3·1 m
3. 100 metres — Dave 12·3 sec, Eric 12·18 sec
4. 100 metres — Sonia 13·04 sec, Tina 13·2 sec
5. Shot put — Ian 9·09 m, Jack 9·1 m

E Copy and complete the table.

DECIMAL FRACTION	IN WORDS	COMMON FRACTION	IN WORDS
7·3	**seven point three**	$\mathbf{7\frac{3}{10}}$	**seven and three tenths**
5·9			
3·47	**three point four seven**	$\mathbf{3\frac{47}{100}}$	**three and forty-seven hundredths**
		$\frac{9}{10}$	
	two point five three		
			five and a half
10·09			
		$1\frac{1}{4}$	
	nought point nine three		
6·25			
			three and one fifth
		$\frac{3}{100}$	
0·07			
		$\frac{3}{4}$	

F Look at these signs: > means IS GREATER THAN
< means IS LESS THAN.

Put the correct sign between these pairs.

1. 0·4 > 0·2 **2.** 0·3 □ 0·8 **3.** 0·5 □ 0·9
4. 0·06 □ 0·6 **5.** 0·7 □ 0·07 **6.** 2·0 □ 0·7
7. 0·15 □ 0·51 **8.** 0·6 □ 0·09 **9.** 10·2 □ 9·5
10. 0·07 □ 0·1 **11.** 0·4 □ 0·15 **12.** 0·9 □ 0·85

G Place these in order, the largest first.

1. 0·2, 20, 2, 0·02, 200
2. 15, 0·15, 150, 1·5, 1500

Adding decimals

A

h	t	1's	$\frac{1}{10}$'s	$\frac{1}{100}$'s
	2	6 ·	2	
	0	0 ·	0	3
	8	4 ·	0	7
1	**1**	**0 ·**	**3**	**0**

Add 26·2, 0·03, 84·07
Make sure the decimal points are written one beneath the other.
Think about the positional value of each figure.
Place it in the correct column.

Add these:

1. 2·46 + 1·67

2. 3·25 + 4·65 + 7·08

3. 4·7 + 3·92 + 0·04

4. 6·37 + 17·02 + 4·08

5. 0·42 + 7·63 + 2·05

6. 4·44 + 6·01 + 0·55

7. 0·01 + 4·0 + 16·0

8. 5·75 + 4·06 + 8·22

9. 2·5 + 25·02 + 10·4 + 7·08
10. 7 + 0·04 + 3·9
11. 0·06 + 0·05 + 0·9
12. 30 + 10·7 + 4·01
13. 0·07 + 5·1 + 6·09
14. 47·02 + 32·8 + 20·18
15. 0·05 + 2 + 0·3
16. 1·09 + 0·9 + 0·07
17. 210 + 0·3 + 1·9

B Six teams ran in the 4 by 100 metres relay race.
The time for each leg was recorded to 0·01 of a second.
Work out the total time for each team. Who won?

TEAM	A	B	C	D	E	F
1st leg – seconds	12·02	12·5	11·98	11·53	12·17	12·8
2nd leg – seconds	12·34	11·97	12·07	12·86	11·77	11·59
3rd leg – seconds	12·09	11·88	12·41	12·34	11·9	11·88
4th leg – seconds	11·57	11·6	11·85	11·98	11·59	11·7
TOTAL TIME – seconds	**48·02**					

C Work out the perimeter of each shape.

1.

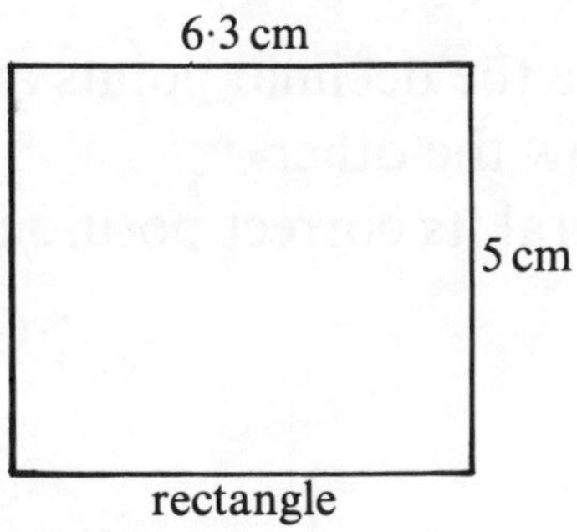

2.

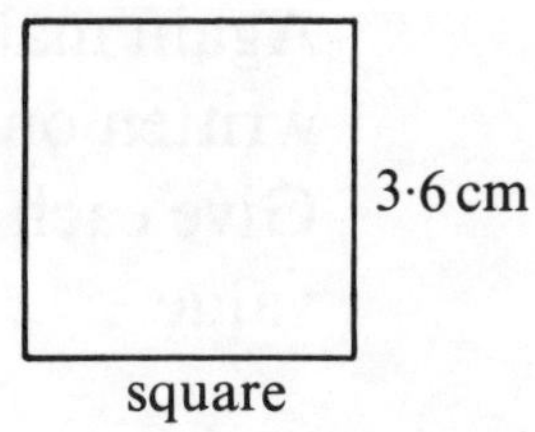

3.

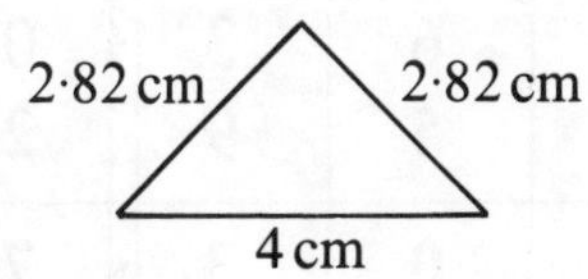

4.

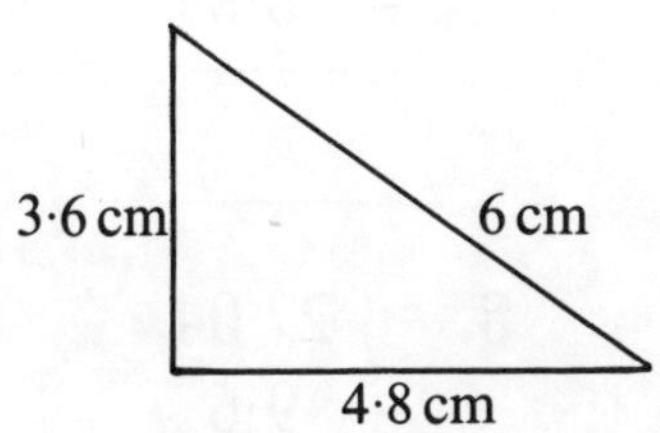

5.

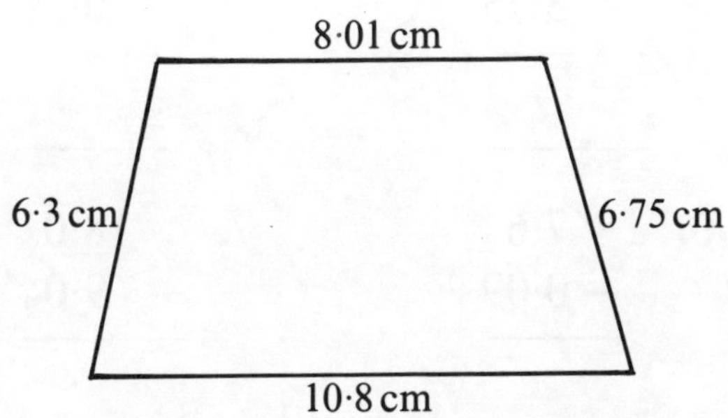

6.

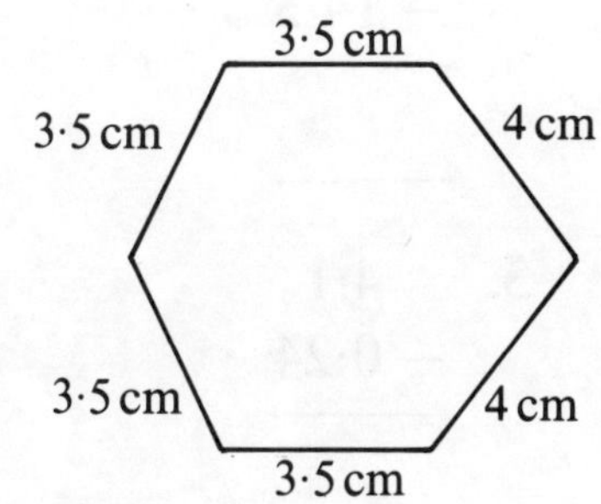

D There are 6 ways to travel from Wick to Zons by way of Voss.

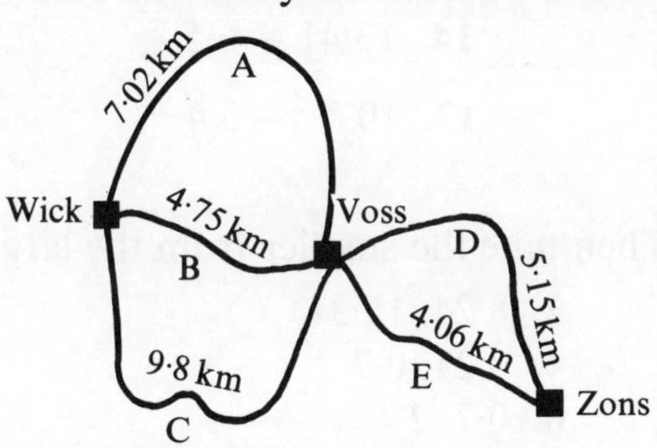

Work out the distances going by the roads:

A → D = (**7·02 km + 5·15 km**) = □

B → D = (   ) = □

C → D = (   ) = □

A → E = (   ) = □

B → E = (   ) = □

C → E = (   ) = □

Which was the shortest route?

E The rain which fell each day of a week was: 0·54 cm, 0·6 cm, 0·25 cm, 0·05 cm, 0·8 cm, 0·15 cm, 0·09 cm.
Work out the total rainfall for the week.

Subtracting decimals

A

h	t	1's	$\frac{1}{10}$'s	$\frac{1}{100}$'s
1	6	3 ·	0	3
	5	9 ·	2	4
1	**0**	**3 ·**	**7**	**9**

163·03 − 59·24
Again make sure the decimal points are written one below the other.
Give each numeral its correct positional value.

Work out these differences:

1. 26·4 − 14·3 = __·__

2. 17·25 − 3·75 = __·__

3. 4·05 − 0·09 = __·__

4. 15·87 − 8·89 = __·__

5. 4·1 − 0·24 = __·__

6. 7·6 − 0·09 = __·__

7. 18·0 − 9·04 = __·__

8. 27·04 − 9·6 = __·__

9. 3·11 − 0·4 **10.** 7·48 − 5·97 **11.** 0·42 − 0·07

12. 0·43 − 0·26 **13.** 0·6 − 0·02 **14.** 13·41 − 6·5

15. 5 − 0·09 **16.** 2 − 0·03 **17.** 10 − 3·4

B

Decide which is the larger. Then take the smaller from the larger.

1. 24·7, 9·07 **2.** 8·24, 15·31
3. 0·09, 0·5 **4.** 0·21, 0·7
5. 1·03, 0·09 **6.** 0·7, 2
7. 10, 15·2 **8.** 10, 1·52

C

Mark walked along 2 sides of the playground, 50·75 m and 46·8 m.
Ruth walked across diagonally, 72·18 m.

1. How far did Mark walk altogether?
2. How much shorter was Ruth's route than Mark's?
3. Mark went back by Ruth's way. How far did he walk altogether?

D The meter reader went from house to house recording the electricity used. His book looked like this:

	PRESENT READING	PREVIOUS READING
1.	3509·4	3014·7
2.	3908·6	3399·4
3.	4260·3	3906·8
4.	4480·0	4260·3
5.	5002·6	4480·0

Work out the number of units of electricity used by each house. Which house had used least? Which had used most?

E These were the kilometre readings on a car's odometer.

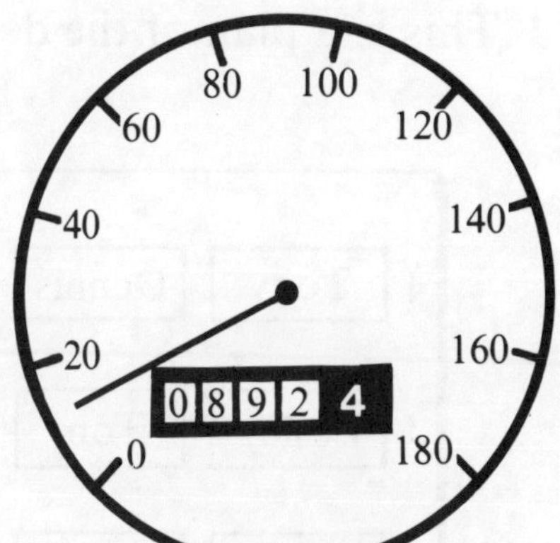

	(start) km	(finish) km
(a)	892·4	1011·6
(b)	1011·6	1427·5
(c)	1427·5	2000·2
(d)	2000·2	2146·0
(e)	2146·0	2424·7

1. Work out the length of each journey.

2. How far had the car gone from the start of the first journey to the end of the last?

F

1. On two days a ship sailed 380·8 km and 400·2 km.
   (a) How far did it sail altogether?
   (b) How much farther did the ship sail on the second day than on the first?

2. Nick paid £16.05 for a calculator. How much change did he get from £20?

3. Jane measured the room. She wrote 4·07 m instead of 4·70 m. How much was she out?

Fixing positions

A

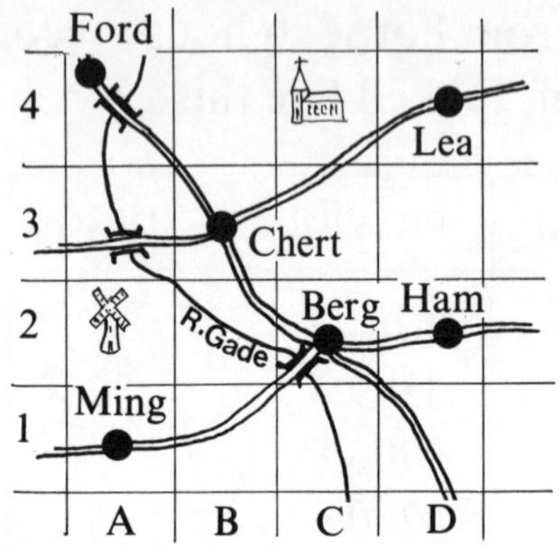

Look at this map.
The position of the village of Lea is (D,4)

1. Give the position of:
   (a) Ford   (b) Chert
   (c) the church   (d) the windmill.

2. What village is at:
   (a) (C,2)   (b) (A,1)   (c) (D,2)?

3. Name the squares through which the River Gade flows.
   Begin with (A,4).

B Desks in a class are often arranged in rows.
This is a plan of the desks in a room.

Row (R)						
4	Tony	Dennis	Margery	Amanda	Ian	Jim
3	Lesley	Tom	Irene	Bill		Edna
2	Nick	Jean	Brenda	Bob	Stan	Ruth
1	Alan	Peggy	Judy	Susan	Lynn	Harry
Column (C)	1	2	3	4	5	6

1. Give the seating position of these children:

	(C,R)		(C,R)		(C,R)		(C,R)
Edna	**(6,3)**	Bob	(,)	Stan	(,)	Lesley	(,)
Peggy	(,)	Irene	(,)	Harry	(,)	Margery	(,)

2. Who sits at: (a) (2,3)   (b) (3,2)   (c) (4,4)   (d) (4,1)
   (e) (1,4)   (f) (5,1)   (g) (1,2)   (h) (2,2)?

3. Which desk is unoccupied?

C The CO-ORDINATES of a point are given by a pair of numbers.

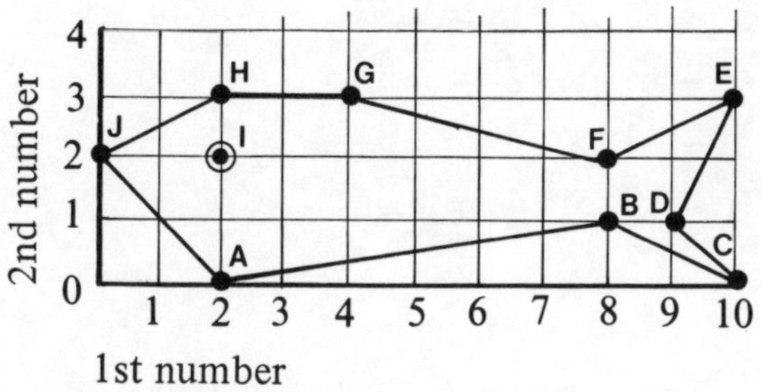

ORDER is important.
The point H is given as (2,3).
F is the point (8,2).
The *horizontal* number comes first.

Write the CO-ORDINATES of the other points.

D

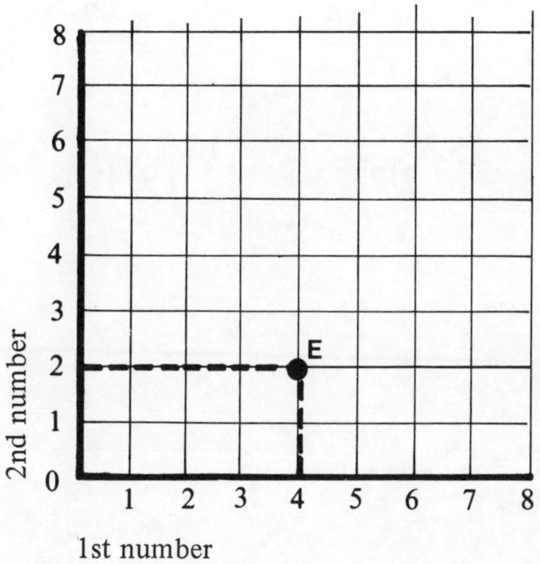

On graph paper marked in centimetre squares, draw 2 perpendicular lines. Write 0 in the corner, and then 1, 2, 3, ... and so on along each line, each number to be 1 cm along.
On your graph mark the points:
A (2,2), B (3,1), C (6,1), D (8,2),
E (4,2), F (4,6), G (4,8).

Join the points A → B → C → D → A
A → F → E
and D → G → E.

E On another pair of horizontal and vertical axes (lines), mark the points:
A (7,1), B (8,3), C (14,3), D (8,10), E (7,14), F (6,10), G (0,3), H (6,3).
Join A → B → C → D → E → F → G → H → A
also H → F and B → D.

F On another pair of axes plot:
A (5,1), B (10,1), C (12,5), D (11,4), E (5,5), F (3,7), G (1,6), H (3,5), I (3,6).
Join A → B → C → D → E → F → G → H → A.

Checking up

A Work out and continue both sequences.

1. $2 \times 2 - 1 \times 3 = \square$
$3 \times 3 - 2 \times 4 = \square$
$4 \times 4 - 3 \times 5 = \square$
$5 \times 5 - 4 \times 6 = \square$

2. $4 \times 3 - 2 \times 1 = \square$
$5 \times 4 - 3 \times 2 = \square$
$6 \times 5 - 4 \times 3 = \square$
$7 \times 6 - 5 \times 4 = \square$

B Work out:

1. 259 + 27 + 62 **2.** 15 + 105 + 7 **3.** 579 + 63 + 170

4. 267 − 189 **5.** 360 − 58 **6.** 729 − 48

7. $\begin{array}{r} 25 \\ \times\ 3 \\ \hline \\ \hline \end{array}$ **8.** $\begin{array}{r} 139 \\ \times\ 7 \\ \hline \\ \hline \end{array}$ **9.** $\begin{array}{r} 256 \\ \times\ 3 \\ \hline \\ \hline \end{array}$ **10.** $\begin{array}{r} 307 \\ \times\ 6 \\ \hline \\ \hline \end{array}$ **11.** $\begin{array}{r} 98 \\ \times\ 7 \\ \hline \\ \hline \end{array}$

12. $5\overline{)65}$ **13.** $7\overline{)91}$ **14.** $3\overline{)245}$ **15.** $8\overline{)732}$ **16.** $9\overline{)457}$

C Copy, writing the complete sentence:

1. 100 cm is equal to 1 ____ and 1 cm = 0·01 ____.
2. □ mm is equal to 1 cm and 1 mm = 0·□ cm.
3. □ p is equal to £1 and 1p = £□·□□.
4. 60 seconds is equal to 1 ____ and 30 seconds = □ ____.
5. □ days is equal to 1 week and 1 day = □ week.

D Here is a recipe for making bread:

500 g flour 50 g lard 4 g sugar
4 g salt 8 g yeast $\frac{1}{4}$ litre water

1. Find the weight of the dry ingredients.
2. What weight of each ingredient is needed for:
   (a) twice as much bread? (b) half as much bread?

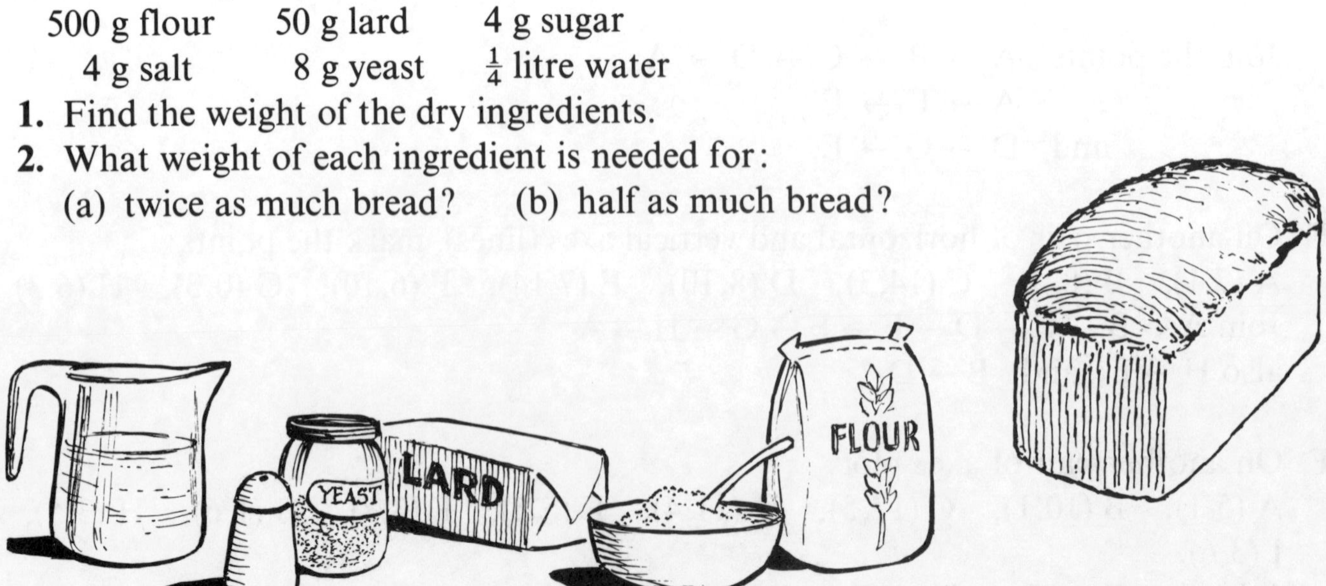

E Work out the area of the rectangles.

1.

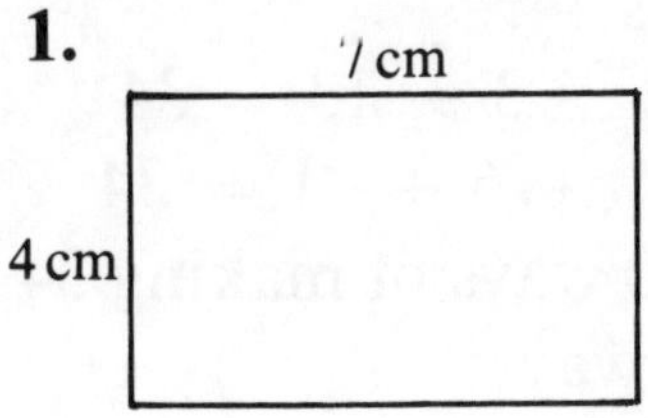

2.

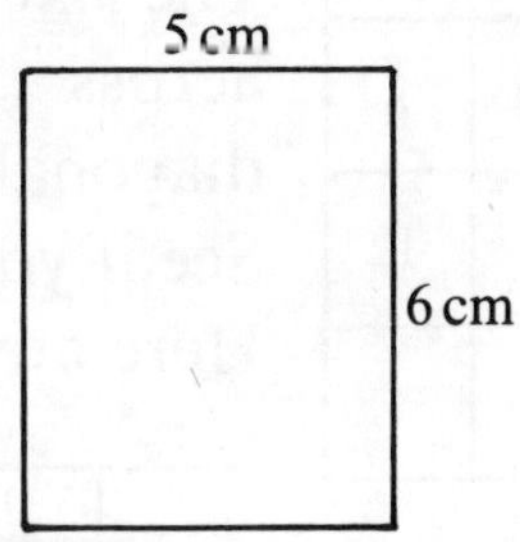

3.

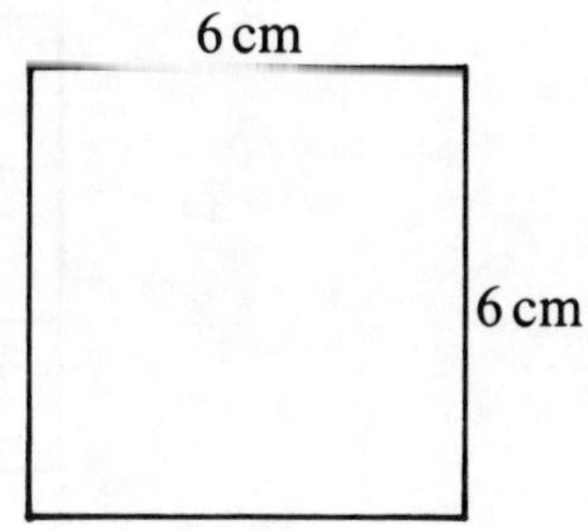

Which one is a square?

F **1.** Change each fraction to its lowest form.

(a) $\frac{4}{20} = \frac{\square}{\square}$ (b) $\frac{5}{25} = \frac{\square}{\square}$ (c) $\frac{5}{50} = \frac{\square}{\square}$

(d) $\frac{14}{40} = \frac{\square}{\square}$ (e) $\frac{10}{20} = \frac{\square}{\square}$

2. Write each of the above fractions as hundredths $\rightarrow \frac{4}{20} = \frac{20}{100}$.

3. Write each as a decimal fraction.

G Work out:

1. $\frac{1}{2} + \frac{1}{4}$ **2.** $\frac{1}{3} + \frac{1}{6}$ **3.** $\frac{3}{5} + \frac{1}{10}$ **4.** $\frac{1}{4} + \frac{3}{8}$ **5.** $\frac{5}{12} + \frac{1}{6}$

6. $\frac{7}{8} - \frac{1}{2}$ **7.** $\frac{5}{6} - \frac{1}{3}$ **8.** $\frac{7}{12} - \frac{1}{4}$ **9.** $\frac{9}{10} - \frac{4}{5}$ **10.** $\frac{3}{4} - \frac{3}{8}$

H

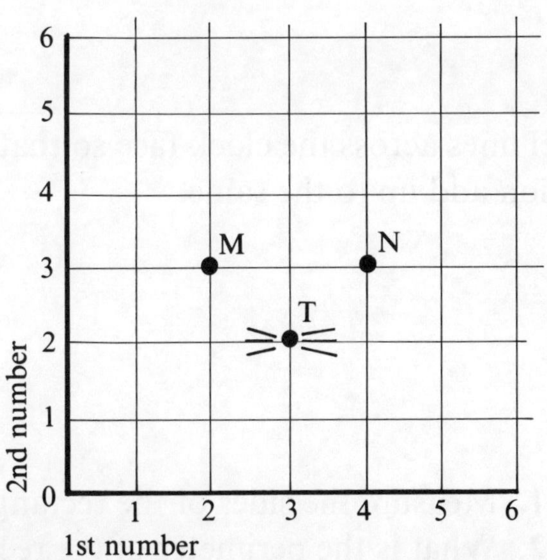

Copy the pair of axes on 1 cm squared paper.

1. Write the co-ordinates (position) of the points in the diagram: M(,), N(,), T(,).

2. Mark these points on your graph.

3. Mark the points: A(1,2), B(2,1), C(4,1), D(5,2), E(5,5), F(4,4), G(2,4), H(1,5).

Join A → B → C → D → E → F → G → H → A.

I Work out:

1. 23·04 + 1·5 + 0·04 **2.** 7·08 + 0·92 + 0·66 **3.** 10·72 + 4·55 + 3·6

4. 68·7 − 30·11 **5.** 54·28 − 6·3 **6.** 18·1 − 8·72

Some puzzles

A

16	2	3	13
5	11	10	8
9	7	6	12
4	14	15	1

This really is a MAGIC SQUARE.
The MAGIC NUMBER is 34.
across $\rightarrow 16 + 2 + 3 + 13 = 34$
diagonally$\rightarrow$ $16 + 11 + 6 + 1 = 34$
See if you can find 20 ways of making 34.
Here are some more ways:

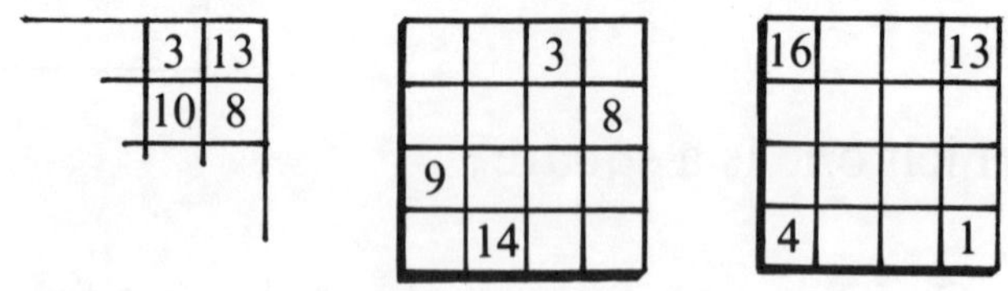

B

With 3 matches you can make an equilateral triangle.
Make 5 equilateral triangles with 9 matches.
How many equilateral triangles can you make with 18 matches?

C

Usually we multiply the top and bottom number of a fraction by the same number to keep it equivalent $\rightarrow \frac{1}{2} \times \frac{2}{2} = \frac{2}{4} = \frac{1}{2}$.
What can you ADD to the top and bottom of $\frac{1}{3}$ so that it is doubled?

D

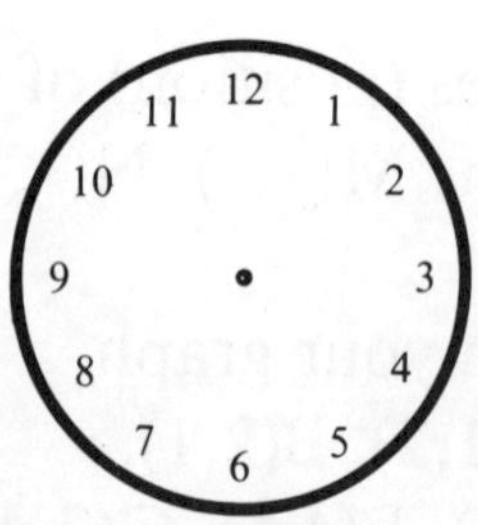

Draw a pair of parallel lines across the clock face so that the numbers in each division add up to the same.

E

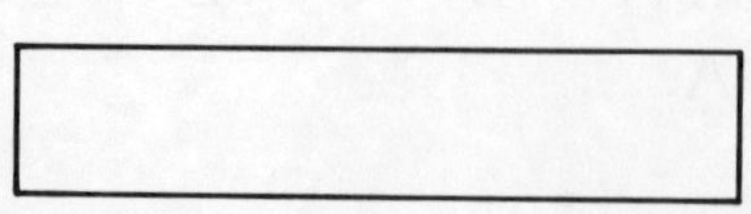

1. Measure the sides of the rectangle.
2. What is the perimeter of the rectangle?
3. How many more rectangles like this are needed to make a square?
4. Work out the area of the square.

Answers

PAGE 6

A $6 \times 6 = 36$
$7 \times 7 = 49$
$8 \times 8 = 64$

B **1.** $25 = 5 \times 5$ **2.** $36 = 6 \times 6$ **3.** $49 = 7 \times 7$

C {*1, 4, 9, 16, 25,* 36, 49, 64, 81, *100*}

D **1.** $11 \times 11 = 121$ **2.** $12 \times 12 = 144$
3. $13 \times 13 = 169$

PAGE 7

E 21, 28
1. 10 **2.** 15 **3.** 21 **4.** 28 **5.** $\cdots + 8 = 36$
6. $\cdots + 8 + 9 = 45$ **7.** $\cdots + 8 + 9 + 10 = 55$

F {*1, 3, 6, 10, 15,* 21, 28, 36, 45, 55}

G **1.** $\begin{cases} 15 + 21 = 36 \\ 6 \times 6 = 36 \end{cases}$ **2.** $\begin{cases} 21 + 28 = 49 \\ 7 \times 7 = 49 \end{cases}$

PAGE 8

A

	thousands	hundreds	tens	ones
1.	5	2	4	6
2.	4	0	7	0
3.	9	0	0	6
4.	6	2	0	0
5.	7	0	8	8

B **1.** 3152 **2.** 7030 **3.** 2400 **4.** 350 **5.** 4006

C **1.** 6 thousand 5 hundred 2 tens 0 ones
2. 5 thousand 0 hundred 4 tens 4 ones
3. 1 thousand 2 hundred 5 tens 6 ones
4. 6 thousand 2 hundred 0 tens 6 ones
5. 9 thousand 4 hundred 7 tens 1 one

D **1.** 7321 **2.** 5200 **3.** 5330 **4.** 9741 **5.** 8644
6. 5432 **7.** 6431 **8.** 9742 **9.** 7420

PAGE 9

E 9999

F **1.** 998, 999, 1000, 1001, 1002
2. 1997, 1998, 1999, 2000, 2001
3. 2499, 2500, 2501, 2502
4. 3000, 3001, 3002, 3003, 3004, 3005, 3006, 3007, 3008, 3009, 3010
5. 3002, 3001, 3000, 2999, 2998, 2997
6. 1000, 999, 998, 997, 996, 995
7. 2703, 2702, 2701, 2700, 2699
8. 3009, 3008, 3007, 3006, 3005, 3004, 3003, 3002, 3001, 3000, 2999, 2998

G **1.** 1001 **2.** 3000 **3.** 3002 **4.** 6010 **5.** 4998
6. 4998 **7.** 4998 **8.** 1990

H R. Amazon
R. Nile
R. Mississippi
R. Volga
R. Danube
R. Rhine

I 1st – N. Parker
2nd – J. Bull
3rd – C. Side
4th – A. Lamb
5th – T. Mann
6th – B. Rush

J

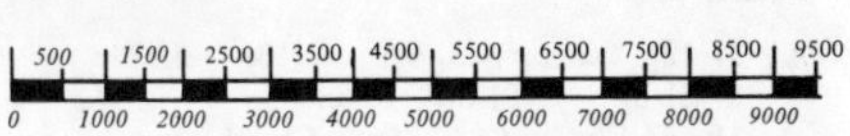

1. 1000 **2.** 2000 **3.** 3000 **4.** 5000 **5.** 7000
6. 10 000

PAGE 10

A **1.** 129 **2.** 120 **3.** 154 **4.** 155 **5.** 120
6. 259 **7.** 322 **8.** 206 **9.** 307 **10.** 410
11. 200 **12.** 403 **13.** 881 **14.** 830 **15.** 806
16. 812 **17.** 748 **18.** 930 **19.** 903 **20.** 861
21. 612 **22.** 899 **23.** 1100 **24.** 1009
25. 1110 **26.** 1057 **27.** 1600 **28.** 1000
29. 1998 **30.** 1554

B **1.** 109 **2.** 93 **3.** 566 **4.** 1013 **5.** 667
6. 528 **7.** 618 **8.** 872 **9.** 644 **10.** 478

PAGE 11

C **1.** Week 1 – 257 km
Week 2 – 199 km
Week 3 – 196 km
2. Mondays – 134 km
Tuesdays – 29 km
Wednesdays – 114 km
Thursdays – 349 km
Fridays – 26 km
3. 652 km overall
4. Thursday Week 1
5. Week 1

D **1.** 302 km **2.** 318 km **3.** 361 km **4.** 368 km
5. 190 km **6.** 615 km

E **1.** 1st Year – 57
2nd Year – 59
3rd Year – 62
4th Year – 58
2. 236 children

PAGE 12

A **1.** 405 **2.** 583 **3.** 38 **4.** 177 **5.** 184 **6.** 89
7. 68 **8.** 307 **9.** 51 **10.** 227 **11.** 481
12. 272 **13.** 665 **14.** 172 **15.** 631 **16.** 516
17. 309 **18.** 605 **19.** 326 **20.** 501 **21.** 156
22. 254 **23.** 492 **24.** 356 **25.** 329 **26.** 859

PAGE 13

B **1.** 595 **2.** 348 **3.** 122 **4.** 81 **5.** 500 **6.** 248
7. 68 **8.** 277 **9.** 248

C **1.** 126 **2.** 70 **3.** 173 **4.** 72 **5.** 95 **6.** 507
7. 170 **8.** 216

D **1.** 83 **2.** 340 **3.** 243 **4.** 189 **5.** 398 **6.** 52
7. 148 **8.** 211 **9.** 49 **10.** 378

E all answers → 1089

PAGE 14

A **1.** *4*, 4 } → 0; 4, 4 } → 0

2. *8*, 16 } → 8; 16, 24 } → 8; 24, 32 } → 8; (8, 8) } → 0; (8, 8) } → 0

B

■	[1] 2	[2] 3	■	■
[3] 7	■	[4] 4	[5] 5	■
[6] 2	[7] 6	■	[8] 6	[9] 2
■	[10] 5	[11] 4	■	7
■	■	[12] 3	2	■

C 15

1.

5	*10*	*3*
4	6	8
9	*2*	*7*

2.

10	20	*6*
8	*12*	16
18	4	*14*

3.

24	3	18
9	15	21
12	27	6

4.

11	21	*7*
9	13	*17*
19	5	15

PAGE 15

D

SCHOOL						
boys girls	*394* *427*	*292* *318*	*426* 434	*339* *349*	*245* 255	*592* *624*
TOTAL	821	610	*860*	*688*	*500*	1216

E **1.** 636 **2.** 643 **3.** Australia, by 16 runs
4. England, 9 more runs **5.** Australia, by 7 runs
6. 1279

F

FIELD	1	2	3	4
side a	*240* m	*194* m	*218* m	*156* m
side b	*220* m	*245* m	*239* m	280 m
side c	320 m	320 m	323 m	323 m
side d	*210* m	231 m	*210* m	231 m

PAGE 16

A

cm	1	2	3	4	5	6	7	8	9
m	*0·01*	*0·02*	0·03	0·04	0·05	0·06	0·07	0·08	0·09

cm	10	11	12	13	14	15	16	17	18	19
m	*0·10*	*0·11*	0·12	0·13	0·14	0·15	0·16	0·17	0·18	0·19

cm	20	30	40	50	60	70	80	90	100
m	*0·20*	*0·30*	0·40	0·50	0·60	0·70	0·80	0·90	1·00

cm	110	120	130	140	150	160	170	180	190	200
m	*1·10*	*1·20*	1·30	1·40	1·50	1·60	1·70	1·80	1·90	2·00

B

CENTIMETRES (cm)	m	cm	METRES (m)
25		*25*	*0·25*
10		*10*	*0·10*
125	*1*	25	*1·25*
235	*2*	*35*	2·35
140	1	40	*1·40*
264	2	64	2·64
205	2	5	*2·05*
309	3	9	3·09
450	*4*	*50*	4·50
7		7	0·07
400	4		4·00
206	*2*	*6*	2·06

PAGE 17

C **1.** 75 cm **2.** 62 cm **3.** 40 cm **4.** 9 cm
5. 7 cm **6.** 5 cm **7.** 8 cm **8.** 2 cm
9. 100 cm **10.** 107 cm **11.** 170 cm
12. 108 cm **13.** 180 cm **14.** 106 cm
15. 160 cm **16.** 275 cm **17.** 472 cm
18. 305 cm **19.** 609 cm **20.** 389 cm

D **1.** *1·25 m* **2.** 2·46 **3.** 1·27 **4.** 1·07 **5.** 2·05
6. 1·50 **7.** 1·86 **8.** 1·10 **9.** 1·01 **10.** 0·53 m
11. 0·74 **12.** 0·60 **13.** 0·06 **14.** 0·40
15. 0·04 **16.** 0·09 **17.** 0·90

E **1.** *1 m 63 cm* **2.** 1 m 48 cm **3.** 1 m 26 cm
4. 2 m 80 cm **5.** 2 m 8 cm **6.** 2 m 0 cm
7. 1 m 60 cm **8.** 1 m 6 cm **9.** 3 m 70 cm
10. 3 m 7 cm **11.** 90 cm **12.** 20 cm **13.** 30 cm
14. 3 cm **15.** 5 cm **16.** 50 cm **17.** 1 m 50 cm

F **1.** metres **2.** centimetres **3.** metres **4.** metres
5. centimetres **6.** centimetres **7.** centimetres

PAGE 18

A **1.** *5 m 26 cm* **2.** 4 m 37 cm **3.** 8 m 90 cm **4.** 6 m 0 cm **5.** 7 m 31 cm **6.** 2 m 13 cm **7.** 4 m 1 cm **8.** 6 m 43 cm **9.** *4·09 m* **10.** 5·50 m **11.** 7·00 m **12.** 14·07 m **13.** 9·80 m **14.** 3·84 m **15.** 3·10 m **16.** 10·00 m

B **1.** (a) 120 cm (b) 1 m 20 cm **2.** (a) 3·04 m (b) 4·27 m **3.** (a) 5·10 m (b) 5 m 10 cm

PAGE 19

A **1.** 1 m 46 cm **2.** 3 m 91 cm **3.** 1 m 28 cm **4.** 1 m 18 cm **5.** 81 cm **6.** 4 m 9 cm **7.** 44 cm **8.** 5 m 36 cm **9.** 2·96 m **10.** 3·08 m **11.** 0·47 m **12.** 2·84 m **13.** 0·75 m **14.** 0·88 m **15.** 1·47 m **16.** 0·04 m

D **1.** (a) 1·60 m (b) 28 cm (c) 16 cm
2. (a) 98 cm (b) 1·57 m (c) 49 cm

PAGE 20

A **1.** . . . *10*
20
30
8 × *4* + 8 = 40
10 × *4* + 10 = 50
12 × *4* + 12 = 60
14 × *4* + 14 = 70

2. . . . *10*
20
30
7 × *6* − *2* = 40
9 × *6* − 4 = 50
11 × *6* − 6 = 60
13 × *6* − 8 = 70

B **1.** . . . *9*
98
987
9876
etc.

2. . . . *8*
86
864
8642
etc.

C **1.** 111
2. 222
3. 333
37 × 12 = 444
37 × 15 = 555
etc.

D **1.** . . . 2
2
2
5 × *6* − *4* × *7* = *2*
6 × *7* − *5* × *8* = *2*
7 × *8* − *6* × *9* = *2*

2. . . . 7
8
9
10
11
12

PAGE 21

E **1.** *14*
21
28
35
42
49
56
63
70

2.

4	*5*	*6*	*8*	*9*
8	10	12	16	18
12	15	18	24	27
16	20	24	32	36
20	25	30	40	45
24	30	36	48	54
28	35	42	56	63
32	40	48	64	72
36	45	54	72	81
40	50	60	80	90

F

10	10	20	30	40	50	60	70	80	90	100
9	9	18	27	36	45	54	63	72	81	90
8	8	16	24	32	*40*	48	56	64	72	80
7	7	14	21	28	35	42	49	56	63	70
6	6	12	18	24	30	36	*42*	48	54	60
5	5	10	*15*	20	25	30	35	40	45	50
4	4	8	12	16	20	24	28	32	36	40
3	3	6	9	12	15	18	21	24	27	30
2	2	*4*	6	8	10	12	14	16	18	20
1	1	2	3	4	5	6	7	8	9	10
×	*1*	*2*	*3*	*4*	*5*	*6*	*7*	*8*	*9*	*10*

1. 15 **2.** 4 **3.** 42 **4.** 40 **5.** 12 **6.** 28 **7.** 28 **8.** 54 **9.** 54 **10.** 25 **11.** 21 **12.** 24 **13.** 56 **14.** 56 **15.** 64 **16.** 18 **17.** 14 **18.** 63 **19.** 63 **20.** 100 **21.** 32 **22.** 72 **23.** 72 **24.** 48 **25.** 16 **26.** 24 **27.** 90

G **1.** 15; 21; 45; 63
2. 8; 48; 32; 16
3. 20; 18; 40; 42

×	ODD	EVEN
ODD	odd	even
EVEN	even	even

PAGE 22

A **1.** 8 **2.** 27 **3.** 64 **4.** 16 **5.** 81 **6.** 125 **7.** 216 **8.** 625 **9.** 343 **10.** 512 **11.** 729 **12.** 256

B **1.** 24 **2.** 60 **3.** 120 **4.** 210 **5.** 336 **6.** 504 **7.** 720 **8.** 0 **9.** 0 **10.** 0 **11.** 315 **12.** 48

C **1.** *2* → *4* → 8 → 16 → 32 → 64
2. *2* → *6* → 18 → 54 → 162
3. 4 → *16* → 64 → 256
4. 5 → *25* → 125 → 625
5. 3 → 9 → *27* → 81 → 243
6. *1* → *10* → 100 → 1000

PAGE 23

D **1.** 108 **2.** 624 **3.** 756 **4.** 441 **5.** 744 **6.** 465 **7.** 574 **8.** 885 **9.** 704 **10.** 753 **11.** 0 **12.** 894 **13.** 1058 **14.** 1776 **15.** 2975

E **1.** 30 **2.** 70 **3.** 40 **4.** 120 **5.** 110 **6.** 170 **7.** 140 **8.** 130 **9.** 210 **10.** 200 **11.** 250 **12.** 300 **13.** 380 **14.** 330 **15.** 460 **16.** 500 **17.** 640 **18.** 990 **19.** 900 **20.** 1000

F

1.	2.	3.
742	548	954
833	860	1008
994	1268	900
889	916	963
1141	828	981

PAGE 24

A **1.** (a) 5 (b) 10 (c) 20 (d) 50 (e) 100
2. (a) 50p (b) 10p (c) 5p (d) 2p (e) 1p

B **1.** £1 10p **2.** £1 **3.** £1 50p **4.** £1 70p **5.** £1 6p **6.** £1 40p **7.** £2 60p **8.** £2 75p

C **1.** 50p + 10p + 10p + 10p + 10p + 10p
2. 50p + 50p + 10p + 5p + 2p
3. 50p + 50p + 50p + 50p + 10p + 2p + 1p
4. 50p + 50p + 50p + 5p
5. 50p + 50p + 5p + 2p + 1p
6. 50p + 50p + 50p + 50p + 5p + 2p
7. 50p + 50p + 50p + 50p + 50p + 10p + 5p
8. 50p + 50p + 10p + 1p
or other appropriate combinations.

D **1.** 50p + 20p + 2p
2. 50p + 20p + 20p + 5p + 2p + 1p
3. 50p + 50p + 5p (or £1 + 5p when available)
4. 50p + 50p + 20p + 20p + 5p (or £1 + 20p + 20p + 5p)
5. 50p + 50p + 50p + 50p + 10p + 5p (or £1 + £1 + 10p + 5p)
6. 50p + 50p + 50p + 50p + 50p + 20p + 5p (or £1 + £1 + 50p + 20p + 5p)
7. 50p + 50p + 50p + 50p + 50p + 50p + 5p + 1p (or £1 + £1 + £1 + 5p + 1p)
8. 50p + 50p + 50p + 50p + 50p + 50p + 50p + 5p + 1p (or £1 + £1 + £1 + 50p + 5p + 1p)

PAGE 25

E **1.** £2 – 22 **2.** £1 – 78 **3.** £7 – 07 **4.** £6 – 50

F

pence	£	p	£
125	*1*	*25*	*1·25*
105	1	5	*1·05*
172	*1*	*72*	1·72
206	2	6	*2·06*
417	4	17	4·17
305	*3*	*5*	3·05
74		74	0·74
209	*2*	*9*	2·09

PAGE 26

A **1.** 90p **2.** 74p **3.** 81p **4.** 112p **5.** 90p **6.** £1 17p **7.** £1 11p **8.** £1 0p **9.** £1 87p **10.** £1 6p **11.** £1 75p **12.** £3 18p **13.** £2 28p **14.** £4 1p **15.** £5 0p **16.** £4 67p **17.** £2 20p **18.** £7 0p **19.** £5 13p **20.** £10 0p

B **1.** (a) £4 39p
(b) £19 38p
2. (a) £8 50p
(b) £18 25p
3. (a) £2 90p
(b) £4 35p

PAGE 27

C **1.** £2·98 **2.** £3·03 **3.** £5·81 **4.** £5·35 **5.** £1·61 **6.** £3·82 **7.** £10·00 **8.** £8·50 **9.** £8·10 **10.** £4·63 **11.** £1·22 **12.** £8·00 **13.** £9·01 **14.** £7·00 **15.** £2·08 **16.** £10·00

D **1.** £18·57
2.

SELLING PRICE
£5·40
£1·67
£5·00
£13·04

PAGE 28

A **1.** 12p **2.** 5p **3.** 9p **4.** 13p **5.** 34p **6.** £1 8p **7.** £1 6p **8.** £1 55p **9.** £1 1p **10.** £3 12p **11.** £2 16p **12.** £1 95p **13.** £1 29p **14.** £4 15p **15.** £2 85p **16.** £4 30p **17.** £2 61p **18.** £2 78p **19.** 36p **20.** £7 38p

B **1.** (a) 57p (b) 43p
2. (a) £8 72p (b) £10 90p (c) £9 10p

C **1.** 99p **2.** 90p **3.** 95p **4.** 90p **5.** 50p

PAGE 29

D **1.** £2·14 **2.** £3·24 **3.** £1·16 **4.** £5·17
5. £1·92 **6.** £4·78 **7.** £4·39 **8.** £3·05
9. £4·63 **10.** £4·83 **11.** £4·85 **12.** £1·83
13. £3·78 **14.** £6·57 **15.** £3·58 **16.** £0·83
17. £8·82 **18.** £4·83 **19.** £6·17 **20.** £0·89
21. £3·39 **22.** £5·46 **23.** £6·74 **24.** £1·91

E **1.** £4·76 **2.** £2·40 **3.** (a) £4·60 (b) £5·40
4. £0·91

PAGE 30

A **1.** 8p **2.** 18p **3.** 20p **4.** 42p **5.** 72p
6. 66p **7.** 84p **8.** 50p **9.** 63p **10.** 80p
11. £1 0p **12.** £1 5p **13.** £1 28p **14.** £4 14p
15. £2 70p **16.** £6 0p **17.** £5 0p **18.** £3 10p
19. £6 68p **20.** £25 56p

B **1.**

£	p
3	20
	84
	57
4	61

2.

£	p
10	80
1	28
2	25
14	33

PAGE 31

C **1.** 4p **2.** 4p **3.** 5p **4.** 8p **5.** 6p **6.** 4p
7. 8p **8.** 5p **9.** 8p **10.** 6p **11.** 9p **12.** 8p
13. 9p **14.** 9p **15.** 9p **16.** 7p **17.** 2p
18. 5p **19.** 8p **20.** 7p **21.** 9p **22.** 7p
23. 6p

D **1.** (a) 9p each
(b) 18p, 45p, 63p, 90p
2. (a) 30p
(b) 60p
(c) 90p
(d) 18p
3. (a) £4 each
(b) £8, £20, £36, £40
(c) 6 bottles
4. (a) £6 (b) 4 balls (c) 2 balls

PAGE 32

A (practical work)
1. scalene **2.** isosceles **3.** scalene
4. equilateral **5.** scalene **6.** isosceles
7. isosceles **8.** scalene

B **1.** 13 cm **2.** 17 cm **3.** 9 cm **4.** 15 cm
5. 30 cm **6.** 14 cm **7.** 17 cm **8.** 20 cm

PAGE 33

C (practical work)
Triangle number 5 was right angled.

D Numbers 1, 4 and 5 are isosceles.

E (practical work)

PAGE 34

A **1.** *12 cm* **2.** 20 cm **3.** 32 cm **4.** 60 cm
5. 68 cm **6.** 80 cm **7.** 96 cm

B **1.** 14 cm **2.** 16 cm **3.** 12 cm **4.** 20 cm
5. 34 cm **6.** 44 cm **7.** 62 cm **8.** 74 cm
9. 80 cm **10.** 74 cm
The opposite sides are equal and parallel, and the corners are all right angles.

PAGE 35

C **1.** The corners are not necessarily right angles.
2. These are right angles.
The diagonals are perpendicular to each other.
3. (a) 8 cm (b) 24 cm (c) 28 cm (d) 36 cm
(e) 48 cm (f) 72 cm (g) 76 cm (h) 92 cm

D **1.** The sides are not all equal.
2. They are right angles.
3. (a) 14 cm (b) 18 cm (c) 14 cm
(d) 32 cm (e) 28 cm (f) 44 cm

E **1.** The opposite sides of a kite are not parallel nor equal.
2. (a) 24 cm (b) 42 cm (c) 64 cm

PAGE 36

A **1.** (a) 3 o'clock 3.00
(b) 5 past 5 5.05
(c) half past 10 10.30
(d) quarter to 2 1.45
(e) quarter past 10 10.15
2. (a) 2.55
(b) 5.00
(c) 10.25
(d) 1.40
(e) 10.10
3. (a) 3.10
(b) 5.15
(c) 10.40
(d) 1.55
(e) 10.25

B **1.** 20 minutes **2.** 25 minutes **3.** 20 minutes
4. 10 minutes **5.** 25 minutes **6.** 45 minutes
7. 35 minutes **8.** 50 minutes **9.** 10 minutes
10. 20 minutes **11.** 25 minutes **12.** 45 minutes

C **1.** *11.59 p.m.* **2.** 11.55 a.m. **3.** 12.10 a.m.
4. 12.01 p.m. **5.** 11.45 p.m. **6.** 11.30 a.m.

D **1.** 30 minutes **2.** 15 minutes **3.** 45 minutes
4. 20 minutes **5.** 40 minutes **6.** 10 minutes
7. 50 minutes **8.** 120 minutes **9.** 90 minutes
10. 180 minutes **11.** 240 minutes
12. 300 minutes

PAGE 37

E **1.** 12.10 p.m. **2.** 12.20 p.m. **3.** 12.30 p.m.
4. 12.50 p.m. **5.** 1.00 p.m. **6.** 1.05 p.m.
7. 1.15 p.m. **8.** 1.30 p.m. **9.** 1.40 p.m.
10. 1.50 p.m. **11.** 2.00 p.m. **12.** 2.00 p.m.

F **1.** 70 minutes **2.** 85 minutes **3.** 90 minutes
4. 105 minutes **5.** 125 minutes **6.** 140 minutes
7. 195 minutes **8.** 200 minutes **9.** 250 minutes
10. 350 minutes

G **1.** 1 hour 10 minutes **2.** 1 hour 30 minutes
3. 1 hour 40 minutes **4.** 2 hours 5 minutes
5. 2 hours 30 minutes **6.** 3 hours 5 minutes

H **1.** 30 seconds **2.** 15 seconds **3.** 45 seconds
4. 20 seconds **5.** 40 seconds **6.** 90 seconds
7. 120 seconds **8.** 180 seconds

I **1.** (a) 36 beats (b) 54 beats (c) 72 beats
2. (a) 100 min (b) 140 min (c) 180 min
(d) 260 min

PAGE 38

A **1.** *8 rem 1* **2.** *4 rem 2* **3.** 3 rem 1 **4.** 3 rem 3
5. 2 rem 1 **6.** 2 rem 3 **7.** 4 rem 1 **8.** 8 rem 2
9. 4 rem 1 **10.** 4 rem 3 **11.** 4 rem 3
12. 4 rem 5 **13.** 8 rem 1 **14.** 4 rem 3
15. 6 rem 2 **16.** 5 rem 1 **17.** 5 rem 4
18. 4 rem 2 **19.** 8 rem 2 **20.** 5 rem 4
21. 7 rem 1 **22.** 6 rem 4 **23.** 9 rem 1
24. 6 rem 2

B **1.** 8 rem 3 **2.** 34 rem 1 **3.** 14 rem 2
4. 24 rem 1 **5.** 13 rem 1 **6.** 17 rem 2
7. 12 rem 5 **8.** 13 rem 1 **9.** 15 rem 1
10. 25 rem 1 **11.** 32 **12.** 32 **13.** 83 rem 1
14. 17 rem 3 **15.** 30 rem 3 **16.** 23 rem 2
17. 40 rem 3 **18.** 40 rem 1 **19.** 60 rem 4
20. 85 rem 3

C **1.** *64 → 32* → 16 → 8 → 4 → 2 → 1
2. *81 → 27* → 9 → 3 → 1

PAGE 39

D **1.** (a) 52 whole weeks
(b) 1 day left over
2. (a) 250 matches altogether
(b) 50
3. (a) 31 km
(b) 62 km, 93 km, 155 km

E

COST OF TICKET	NUMBER OF TICKETS	TOTAL TAKINGS
£3	109	*£327*
£5	81	*£405*
£7	42	*£294*
£4	43	*£172*
£12	*8*	*£96*

PAGE 40

A **1.** 30 cm **2.** 63 cm **3.** 48 cm **4.** 72 cm
5. 28 cm **6.** 72 cm **7.** 105 cm **8.** 108 cm
9. 180 cm **10.** 315 cm **11.** 2 m 15 cm
12. 1 m 68 cm **13.** 4 m 84 cm **14.** 10 m 24 cm
15. 8 m 85 cm **16.** 4·56 m **17.** 6·18 m
18. 2·58 m **19.** 4·80 m **20.** 14·00 m

B **1.** (a) 50 m (b) 100 m (c) 175 m
(d) 250 m
2. (a) 52 m (b) 1 m (c) 4·40 m (d) 5 m
3. (a) 1·70 m (b) 2·55 m (c) 5·95 m
(d) 8·50 m

PAGE 41

C **1.** 2 cm **2.** 7 cm **3.** 9 cm **4.** 9 cm **5.** 7 cm
6. 12 cm **7.** 12 cm **8.** 12 cm **9.** 15 cm
10. 17 cm **11.** 15 m **12.** 13 m **13.** 8 m
14. 14 m **15.** 12 m **16.** 0·18 m **17.** 0·25 m
18. 0·40 m **19.** 4·92 m **20.** 0·17 m **21.** 18 cm
22. 15 cm **23.** 85 cm **24.** 56 cm **25.** 57 cm

D **1.** (a) 8 cm
(b) 16 cm, 24 cm, 40 cm
2. (a) 2 cm
(b) 8 cm, 12 cm, 40 cm
3. (a) 10 strips
(b) 6 strips
(c) 5 strips
(d) 3 strips
4. (a) 4 cm
(b) 12 cm
(c) 18 cm
(d) 26 cm
5. (a) 6 times
(b) 8 times
(c) 12 times
(d) 24 times
6. (a) heavy lorry or tractor
(b) bus or lorry
(c) large car or bicycle
(d) child's bicycle

PAGE 42

A **1.** 20 g + 10 g
2. 50 g + 20 g + 10 g
3. 100 g + 20 g + 10 g
4. 100 g + 20 g
5. 50 g + 10 g
6. 100 g + 50 g + 10 g
7. 100 g + 50 g + 20 g
8. 200 g + 10 g
9. 200 g + 20 g + 10 g
10. 200 g + 50 g + 10 g
11. 200 g + 100 g + 10 g
12. 200 g + 100 g + 20 g + 10 g
13. 200 g + 100 g + 50 g + 10 g
14. 200 g + 100 g + 50 g + 20 g
15. 200 g + 100 g + 50 g + 20 g + 10 g
16. 500 g + 10 g

17. 500 g + 20 g
18. 500 g + 20 g + 10 g
19. 500 g + 50 g + 10 g
20. 500 g + 200 g + 50 g
21. 880 g

B **1.** 100 g − 10 g
2. 100 g + 50 g − 10 g
3. 200 g + 50 g − 10 g
4. 200 g + 100 g − 10 g
5. 200 g + 100 g + 50 g − 10 g
6. 500 g − 100 g
7. 500 g − 10 g
8. 500 g − 50 g
9. 500 g − 20 g
10. 500 g + 50 g − 10 g
11. 500 g + 100 g − 10 g
12. 500 g − 100 g − 10 g
13. 500 g − 50 g − 10 g
14. 500 g + 100 g + 50 g − 10 g
15. 500 g + 200 g + 100 g − 10 g
16. Weigh 10 g of Plasticine. Cut this into two, and weigh each piece against the other. Keep taking from one to give to the other until they balance.

PAGE 43

C

WEIGHT	500 g	200 g	100 g	50 g	20 g	10 g
70 g	*0*	*0*	*0*	*1*	*1*	*0*
110 g	0	0	1	0	0	1
130 g	0	0	1	0	1	1
180 g	0	0	1	1	1	1
220 g	0	1	0	0	1	0
270 g	0	1	0	1	1	0
380 g	0	1	1	1	1	1
620 g	1	0	1	0	1	0
760 g	1	1	0	1	0	1
870 g	1	1	1	1	1	0

D **1.** 380 g
2. 515 g
3. (1) flour 500 g, lard or margarine 250 g, salt 10 g
(2) bread 500 g, sugar 100 g, suet 100 g, dried fruit 200 g, spice 10 g, (eggs 120 g)

PAGE 44

A **1.** 91 g **2.** 101 g **3.** 101 g **4.** 100 g **5.** 108 g **6.** 133 g **7.** 213 g **8.** 221 g **9.** 250 g **10.** 390 g **11.** 381 g **12.** 500 g **13.** 580 g **14.** 700 g **15.** 943 g **16.** 630 g **18.** 605 g **18.** 677 g **19.** 417 g **20.** 565 g **21.** 645 g

B **1.** 425 g
2. (a) 300 g (b) 3000 g
3. (a) 454 g (b) 510 g

C e.g. letters, pencils, packets of sweets.

PAGE 45

D **1.** 41 g **2.** 9 g **3.** 81 g **4.** 36 g **5.** 14 g **6.** 90 g **7.** 75 g **8.** 197 g **9.** 47 g **10.** 207 g **11.** 278 g **12.** 455 g **13.** 545 g **14.** 533 g **15.** 275 g **16.** 78 g **17.** 227 g **18.** 247 g **19.** 359 g **20.** 507 g **21.** 227 g **22.** 276 g **23.** 848 g **24.** 675 g

E **1.** 78 g **2.** (a) 50 g (b) 100 g (c) 290 g **3.** 47 g

PAGE 46

A **1.** 81 g **2.** 100 g **3.** 95 g **4.** 276 g **5.** 222 g **6.** 603 g **7.** 546 g **8.** 343 g **9.** 333 g **10.** 560 g **11.** 500 g **12.** 250 g **13.** 750 g **14.** 750 g **15.** 500 g **16.** 900 g **17.** 672 g **18.** 1125 g **19.** 768 g **20.** 704 g

B **1.** (a) 56 g (b) 140 g (c) 168 g (d) 280 g
2. (a) 70 g (b) 105 g (c) 175 g (d) 252 g (e) 336 g (f) 420 g
3. (a) 224 g (b) 336 g (c) 448 g (d) 504 g (e) 560 g (f) 1120 g

PAGE 47

C **1.** 5 g **2.** 50 g **3.** 7 g **4.** 70 g **5.** 6 g **6.** 60 g **7.** 6 g **8.** 60 g **9.** 9 g **10.** 90 g **11.** 40 g **12.** 30 g **13.** 80 g **14.** 70 g **15.** 80 g **16.** $60\frac{4}{5}$ g **17.** $70\frac{1}{6}$ g **18.** $60\frac{3}{4}$ g **19.** $60\frac{5}{8}$ g **20.** $80\frac{2}{5}$ g

D **1.** (a) $50\frac{4}{9}$ g
(b) $90\frac{4}{5}$ g
2. (a) 9 g
(b) 36 g, 63 g, 81 g
3. (a) 100 g ham, 25 g butter, 200 g cooked chicken, 75 g mushrooms, 250 g pastry
(b) 40 g ham, 10 g butter, 80 g chicken, 30 g mushrooms, 100 g pastry

PAGE 48

A **1.** 12 square units
2. 18 square units
3. 24 square units
4. 42 square units
5. 54 square units
6. 60 square units
7. 120 square units
8. 150 square units
9. 270 square units

B (practical work)

C **1.** 6 cm × 8 cm
4 cm × 12 cm
2. 6 cm × 9 cm
3 cm × 18 cm
3. 7 cm × 8 cm
4 cm × 14 cm
4. 7 cm × 9 cm
3 cm × 21 cm
(or other appropriate answers)

PAGE 49

D (practical work)
1. 20 cm² **2.** 18 cm² **3.** 32 cm² **4.** 35 cm² **5.** 48 cm² **6.** 36 cm² **7.** 36 cm² **8.** 36 cm² **9.** 36 cm² **10.** 16 cm² **11.** 30 cm² **12.** 18 cm²

E **1.** 2 cm **2.** 2 cm **3.** 7 cm **4.** 4 cm **5.** 6 cm **6.** 3 cm **7.** 6 cm **8.** 7 cm

F

LENGTH BREADTH	cm cm²	*7* *5*	9 *7*	*5* 11	9 *6*	*12* *7*	*9* *8*	*16* *4*	14 *2*
AREA	cm²	35	*63*	*55*	*54*	84	72	64	*28*

AREA OF A RECTANGLE
= *L*ENGTH × *B*READTH

G **1.** (a) 10 cm (b) 36 cm
2. (a) 30 cm (b) 25 cm (c) 125 cm²

PAGE 50

A **1.** 3 cm square → 9 cm²
2. 4 cm square → 16 cm²
3. 5 cm square → 25 cm²

B **1.** 36 cm² **2.** 49 cm² **3.** 64 cm² **4.** 81 cm² **5.** 100 cm² **6.** 121 cm²

C **1.** (9 − 3) cm² = 6 cm²
2. (16 − 4) cm² = 12 cm²
3. (25 − 5) cm² = 20 cm²
4. (36 − 6) cm² = 30 cm²
5. (49 − 7) cm² = 42 cm²
6. (81 − 9) cm² = 72 cm²

PAGE 51

D **1.** 4 times **2.** 9 times **3.** 9 times **4.** 4 times **5.** 25 times

E **1.** *2 cm* **2.** 4 cm **3.** 5 cm **4.** 10 cm **5.** 3 cm **6.** 6 cm **7.** 8 cm **8.** 7 cm **9.** 9 cm

F **1.** (a) 7 cm (b) 49 cm²
2. (a) 5 cm (b) 20 cm (c) 30 cm

PAGE 52

A **1.** A → *80°* B → 100° C → 65° D → 40° E → 30° F → 0° G → 15° H → −5° J → −10° K → −20°
2. 30° **3.** hot **4.** very cold **5.** about 3° lower **6.** −30° **7.** 110°

B (practical work)

PAGE 53

C **1.** +60° **2.** −60° **3.** +30° **4.** −40° **5.** +5° **6.** +70° **7.** +15° **8.** −80° **9.** +100° **10.** −120° **11.** +35° **12.** +65° **13.** −100° **14.** +50° **15.** −35° **16.** +5°

D **1.** usually a fall in temperature
2. a fall in temperature
3. the liquid expands and contracts
4. it falls
5. it rises

E **1.** 20°C **2.** 851°C **3.** 447°C **4.** 1298°C **5.** 1500°C

F **1.** about 46°F
2. about 54°F
3. about 70°F
4. about 90°F

PAGE 54

A **1.** $\frac{1}{2}$ **2.** $\frac{1}{6}$ **3.** $\frac{1}{3}$ **4.** $\frac{3}{4}$ **5.** $\frac{2}{3}$ **6.** $\frac{5}{6}$ **7.** 1

B **1.** $\frac{1}{5}$ **2.** $\frac{1}{2}$ **3.** $\frac{3}{5}$ **4.** $\frac{4}{5}$ **5.** $\frac{2}{5}$ **6.** 1

C **1.** $\frac{1}{4}$ **2.** $\frac{3}{4}$ **3.** $\frac{1}{2}$ **4.** $\frac{1}{2}$ **5.** 1 **6.** 1 **7.** 1

D **1.** $\frac{5}{10}$ **2.** $\frac{4}{8}$ **3.** $\frac{6}{12}$ **4.** $\frac{4}{12}$ **5.** $\frac{8}{12}$ **6.** $\frac{2}{8}$ **7.** $\frac{3}{12}$ **8.** $\frac{9}{12}$ **9.** $\frac{6}{8}$ **10.** $\frac{10}{12}$ **11.** $\frac{6}{10}$ **12.** $\frac{4}{10}$ **13.** 1 **14.** 1 **15.** 1

PAGE 55

E **1.** 6 mm **2.** 6 mm **3.** 4 mm **4.** 4 mm **5.** 5 mm **6.** 5 mm **7.** 2 mm **8.** 2 mm **9.** 8 mm **10.** 8 mm

F **1.** 10p **2.** 20p **3.** 20p **4.** 40p **5.** 40p **6.** 50p **7.** 50p **8.** 60p **9.** 60p **10.** 80p

G **1.** 6 minutes
2. 12 minutes
3. 12 minutes
4. 30 minutes
5. 30 minutes
6. 30 minutes
7. 36 minutes
8. 36 minutes
9. 15 minutes
10. 5 minutes
11. 10 minutes
12. 10 minutes
13. 20 minutes
14. 20 minutes
15. 40 minutes

H **1.** £2 **2.** £2 **3.** £4 **4.** 25p **5.** 75p **6.** 75p **7.** £1·50 **8.** £7·50 **9.** £6 **10.** 30p **11.** £1·20 **12.** £3

I **1.** $\frac{1}{5}$ **2.** $\frac{1}{6}$ **3.** $\frac{1}{3}$ **4.** $\frac{1}{3}$ **5.** $\frac{1}{2}$ **6.** $\frac{1}{5}$ **7.** $\frac{1}{6}$ **8.** $\frac{1}{2}$ **9.** $\frac{1}{12}$ **10.** $\frac{1}{12}$ **11.** $\frac{1}{6}$ **12.** $\frac{1}{4}$ **13.** $\frac{1}{4}$ **14.** $\frac{1}{4}$ **15.** $\frac{1}{8}$ **16.** $\frac{1}{3}$ **17.** $\frac{1}{6}$ **18.** $\frac{1}{2}$ **19.** $\frac{5}{12}$ **20.** $\frac{2}{3}$

PAGE 56

A **1.** $\frac{3}{2} \rightarrow 1\frac{1}{2}$
2. $\frac{9}{4} \rightarrow 2\frac{1}{4}$
3. $\frac{4}{3} \rightarrow 1\frac{1}{3}$
4. $\frac{14}{6} \rightarrow 2\frac{2}{6} \rightarrow 2\frac{1}{3}$
5. $\frac{11}{6} \rightarrow 1\frac{5}{6}$
6. $\frac{19}{8} \rightarrow 2\frac{3}{8}$

B

centimetres	*1*	*$1\frac{1}{2}$*	2	$2\frac{1}{2}$	3	*$3\frac{1}{2}$*	4	$4\frac{1}{2}$	5	$5\frac{1}{2}$
$\frac{1}{2}$ centimetres	*2*	3	4	*5*	*6*	7	8	*9*	10	*11*
millimetres	5	*15*	20	25	30	35	*40*	45	50	55

C **1.** *4* **2.** 2 **3.** 3 **4.** 4 **5.** 5 **6.** 3 **7.** 5 **8.** 3 **9.** 8 **10.** 4 **11.** 12 **12.** 2 **13.** 3 **14.** 3 **15.** 7

PAGE 57

D **1.** $\frac{6}{2}, \frac{9}{3}, \frac{12}{4}, \frac{15}{5}, \frac{18}{6}, \frac{21}{7}, \frac{24}{8}, \frac{27}{9}, \frac{30}{10} \ldots$
2. $\frac{8}{2}, \frac{12}{3}, \frac{16}{4}, \frac{20}{5}, \frac{24}{6}, \frac{28}{7}, \frac{32}{8}, \frac{36}{9}, \frac{40}{10} \ldots$
3. $\frac{10}{2}, \frac{15}{3}, \frac{20}{4}, \frac{25}{5}, \frac{30}{6}, \frac{35}{7}, \frac{40}{8}, \frac{45}{9}, \frac{50}{10} \ldots$
4. $\frac{12}{2}, \frac{18}{3}, \frac{24}{4}, \frac{30}{5}, \frac{36}{6}, \frac{42}{7}, \frac{48}{8}, \frac{54}{9}, \frac{60}{10} \ldots$
5. $\frac{14}{2}, \frac{21}{3}, \frac{28}{4}, \frac{35}{5}, \frac{42}{6}, \frac{49}{7}, \frac{56}{8}, \frac{63}{9}, \frac{70}{10} \ldots$
6. $\frac{16}{2}, \frac{24}{3}, \frac{32}{4}, \frac{40}{5}, \frac{48}{6}, \frac{56}{7}, \frac{64}{8}, \frac{72}{9}, \frac{80}{10} \ldots$

E **1.** $\frac{3}{2}$ **2.** $\frac{7}{2}$ **3.** $\frac{9}{4}$ **4.** $\frac{11}{4}$ **5.** $\frac{7}{3}$ **6.** $\frac{27}{8}$ **7.** $\frac{21}{8}$ **8.** $\frac{25}{4}$ **9.** $\frac{17}{3}$ **10.** $\frac{14}{3}$ **11.** $\frac{9}{5}$ **12.** $\frac{22}{7}$ **13.** $\frac{31}{8}$ **14.** $\frac{17}{6}$ **15.** $\frac{25}{6}$ **16.** $\frac{7}{5}$ **17.** $\frac{16}{5}$ **18.** $\frac{61}{10}$ **19.** $\frac{49}{10}$ **20.** $\frac{14}{5}$

F **1.** $1\frac{1}{4}$ **2.** $1\frac{3}{4}$ **3.** $3\frac{1}{3}$ **4.** $2\frac{1}{4}$ **5.** $2\frac{1}{5}$ **6.** $4\frac{1}{2}$ **7.** $1\frac{4}{5}$ **8.** $3\frac{1}{3}$ **9.** $3\frac{1}{4}$ **10.** $3\frac{3}{4}$ **11.** $3\frac{1}{7}$ **12.** $3\frac{1}{2}$ **13.** $1\frac{1}{5}$ **14.** $3\frac{2}{3}$ **15.** $5\frac{1}{5}$ **16.** $4\frac{1}{5}$ **17.** $4\frac{5}{7}$ **18.** $5\frac{1}{2}$ **19.** $4\frac{1}{4}$ **20.** $7\frac{2}{3}$

PAGE 58

A **1.** $\frac{2}{4} = \frac{1}{2}$ **2.** $\frac{2}{2} = 1$ **3.** $\frac{2}{3}$ **4.** $\frac{3}{3} = 1$ **5.** $\frac{4}{8} = \frac{1}{2}$ **6.** $\frac{8}{8} = 1$

B **1.** $\frac{2}{3}$ **2.** $\frac{2}{6} = \frac{1}{3}$ **3.** $\frac{2}{4} = \frac{1}{2}$ **4.** $\frac{4}{5}$ **5.** $\frac{2}{8} = \frac{1}{4}$ **6.** $\frac{2}{10} = \frac{1}{5}$ **7.** $\frac{2}{7}$ **8.** $\frac{6}{8} = \frac{3}{4}$ **9.** $\frac{6}{6} = 1$ **10.** $\frac{3}{5}$ **11.** $\frac{4}{10} = \frac{2}{5}$ **12.** $\frac{4}{8} = \frac{1}{2}$ **13.** $\frac{2}{9}$ **14.** $\frac{8}{10} = \frac{4}{5}$ **15.** $\frac{8}{12} = \frac{2}{3}$ **16.** $\frac{4}{5}$ **17.** $\frac{6}{8} = \frac{3}{4}$ **18.** $\frac{3}{9} = \frac{1}{3}$

PAGE 59

C **1.** $\frac{1}{2}$ hour **2.** 1 hour **3.** $1\frac{1}{4}$ hours **4.** $2\frac{1}{2}$ hours

D **1.** $\frac{2}{4}$, $\frac{2}{8}$, $\frac{6}{8}$, $\frac{5}{10}$, $\frac{3}{12}$, $\frac{3}{6}$ **2.** $\frac{2}{6}$, $\frac{2}{12}$, $\frac{6}{12}$, $\frac{4}{12}$, $\frac{8}{12}$, $\frac{4}{6}$ **3.** $\frac{9}{12}$, $\frac{10}{12}$, $\frac{2}{10}$, $\frac{6}{10}$, $\frac{8}{10}$, $\frac{6}{9}$

E **1.** $\frac{3}{8}$ **2.** $\frac{5}{8}$ **3.** $\frac{3}{10}$ **4.** $\frac{4}{6} = \frac{2}{3}$ **5.** $\frac{4}{12} = \frac{1}{3}$ **6.** $\frac{6}{10} = \frac{3}{5}$ **7.** $\frac{5}{12}$ **8.** $\frac{5}{16}$ **9.** $\frac{9}{16}$ **10.** $\frac{3}{6} = \frac{1}{2}$ **11.** $\frac{4}{9}$ **12.** $\frac{6}{20} = \frac{3}{10}$ **13.** $\frac{5}{8}$ **14.** $\frac{5}{10} = \frac{1}{2}$ **15.** $\frac{7}{8}$ **16.** $\frac{11}{12}$ **17.** $\frac{5}{10} = \frac{1}{2}$ **18.** $\frac{8}{10} = \frac{4}{5}$ **19.** $\frac{7}{8}$ **20.** $\frac{11}{12}$ **21.** $\frac{5}{6}$ **22.** $\frac{7}{16}$ **23.** $\frac{7}{8}$ **24.** $\frac{9}{12} = \frac{3}{4}$ **25.** $\frac{5}{8}$

F

TOTAL TIME IN HOURS	$\frac{3}{4}$	$\frac{3}{5}$	$\frac{1}{2}$	$\frac{1}{3}$	$\frac{1}{2}$	$\frac{4}{5}$	$\frac{2}{5}$	$\frac{3}{4}$
TOTAL TIME IN MINUTES	*45*	36	30	20	30	48	24	45

PAGE 60

A **1.** $\frac{2}{4} = \frac{1}{2}$
2. $1 \rightarrow \frac{2}{2} - \frac{1}{2} = \frac{1}{2}$
3. $\frac{1}{3}$
4. $1 \rightarrow \frac{3}{3} - \frac{2}{3} = \frac{1}{3}$
5. $\frac{2}{8} = \frac{1}{4}$
6. $\frac{4}{8} = \frac{1}{2}$

B **1.** $\frac{4}{6} = \frac{2}{3}$ **2.** $\frac{6}{8} = \frac{3}{4}$ **3.** $\frac{2}{12} = \frac{1}{6}$ **4.** $\frac{3}{5}$ **5.** $\frac{6}{10} = \frac{3}{5}$ **6.** $\frac{2}{8} = \frac{1}{4}$ **7.** $\frac{2}{5}$ **8.** $\frac{4}{10} = \frac{2}{5}$ **9.** $\frac{1}{5}$ **10.** $\frac{10}{12} = \frac{5}{6}$ **11.** $\frac{6}{10} = \frac{3}{5}$ **12.** $\frac{2}{8} = \frac{1}{4}$ **13.** $\frac{8}{10} = \frac{4}{5}$ **14.** $\frac{2}{12} = \frac{1}{6}$ **15.** $\frac{6}{10} = \frac{3}{5}$ **16.** $\frac{4}{12} = \frac{1}{3}$

C **1.** $\frac{1}{2}$ **2.** $\frac{2}{3}$ **3.** $\frac{3}{4}$ **4.** $\frac{4}{5}$ **5.** $\frac{7}{8}$ **6.** $\frac{5}{8}$ **7.** $\frac{1}{6}$ **8.** $\frac{5}{12}$ **9.** $\frac{7}{10}$ **10.** $\frac{1}{8}$ **11.** $\frac{1}{3}$ **12.** $\frac{1}{4}$ **13.** $\frac{5}{6}$ **14.** $\frac{11}{12}$ **15.** $\frac{3}{10}$

PAGE 61

D **1.**

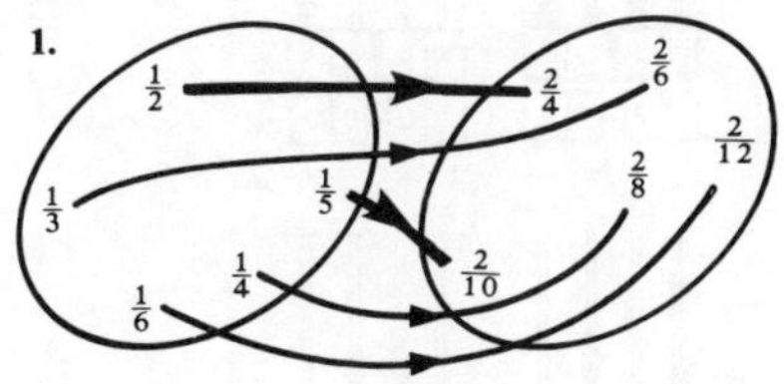

2.

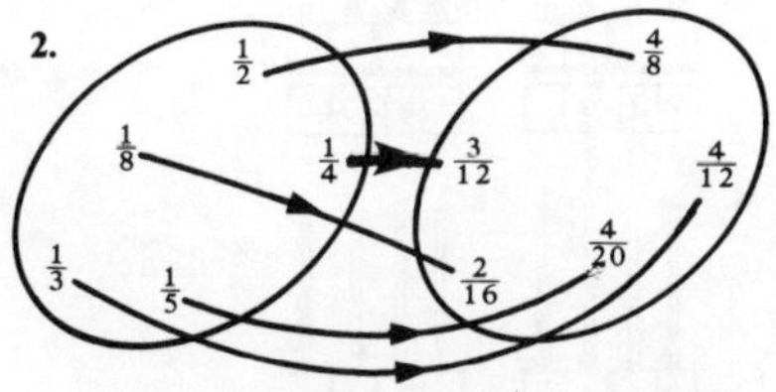

3.

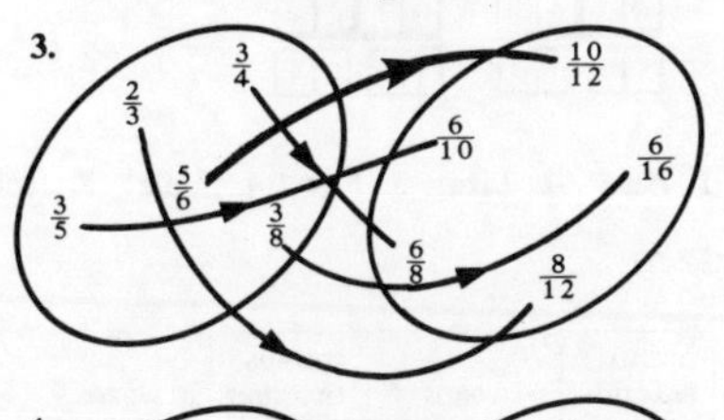

4.

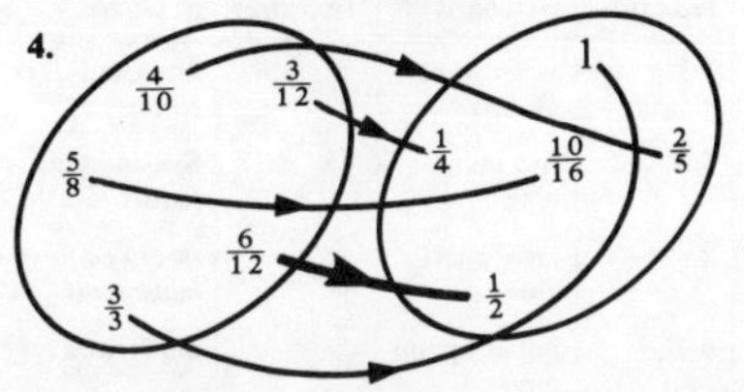

E **1.** $\frac{1}{4}$ **2.** $\frac{1}{3}$ **3.** $\frac{2}{5}$ **4.** $\frac{3}{8}$ **5.** $\frac{5}{12}$ **6.** $\frac{1}{8}$ **7.** $\frac{1}{4}$ **8.** $\frac{1}{10}$ **9.** $\frac{1}{6}$ **10.** $\frac{1}{6}$ **11.** $\frac{1}{4}$ **12.** $\frac{3}{8}$ **13.** $\frac{9}{16}$ **14.** $\frac{1}{5}$ **15.** $\frac{1}{2}$ **16.** $\frac{1}{10}$ **17.** $\frac{1}{3}$ **18.** $\frac{1}{8}$ **19.** $\frac{3}{10}$ **20.** $\frac{1}{8}$ **21.** $\frac{1}{6}$ **22.** $\frac{3}{8}$ **23.** $\frac{1}{10}$ **24.** $\frac{3}{10}$ **25.** $\frac{1}{3}$

F $\frac{3}{8}$ km still to go
$\frac{5}{8}$ km already covered

PAGE 62

A (individual lists)
B 1, 3, 4, 6 and 8 are perpendicular

PAGE 63

C **1.**
2.
3.

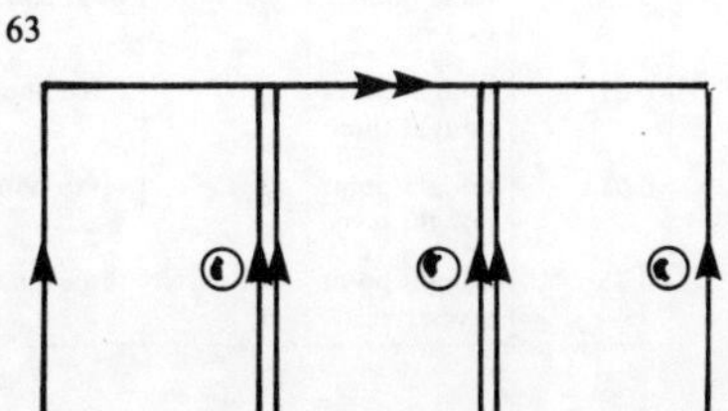

D

SHAPE	NUMBER OF PAIRS OF PERPENDICULAR LINES	NUMBER OF PAIRS OF PARALLEL LINES
1	*2*	*1*
2	4	2
3	0	2
4	1	2
5	2	1
6	5	2
7	1	0
8	0	4

PAGE 64

A **1.** → S
2. → W
3. → N

B **1.** E **2.** N **3.** W **4.** S **5.** W **6.** S **7.** E **8.** N

C **1.** N **2.** S **3.** W **4.** E **5.** N **6.** S **7.** W **8.** E
9. They have the same result.

D **1.** N **2.** W **3.** E **4.** S **5.** S **6.** E **7.** W **8.** N

PAGE 65

E **1.** *E → S*
2. S → W
3. S → N
4. S → E
5. S → W
6. S → E
7. E → N
8. W → E
9. W → S
10.

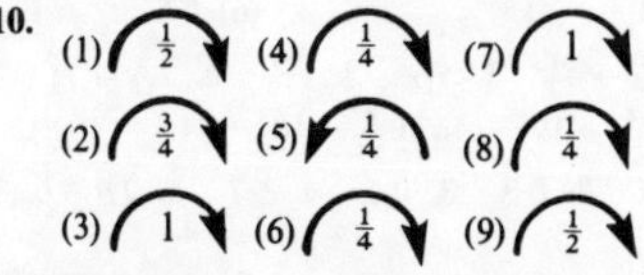

F **1.** S **2.** N **3.** W **4.** W **5.** N **6.** S **7.** N

G **1.** 1 RIGHT ANGLE is $\frac{1}{4}$ of a full circle.
2. 2 RIGHT ANGLES make $\frac{1}{2}$ of a full circle.
3. 3 RIGHT ANGLES make $\frac{3}{4}$ of a full circle.

PAGE 66

A

		RUNS SCORED
J. Hobbs		*13*
R. Ginn		22
P. Knights		9
F. Day		29
D. Pound		0
S. Pence		7
J. Hunt		11
C. Ball		15
A. Cross		16
F. Patch		8
D. Amuah		7
EXTRAS		7
TOTAL SCORE		144

F. Day made the highest score.

B **1.**

VEHICLES	
private cars	卌 卌 卌 卌 卌 ǁ
vans	卌 卌 卌 卌 |
lorries	卌 卌 ǁ
motor cycles	卌 卌 卌
bicycles	卌 卌 卌 卌 卌 ǁǁ

2. ·104 vehicles
3. bicycles

PAGE 67

C **1.** *28 rainy days* **2.** 150 passengers **3.** 50 babies **4.** 750 conifers **5.** 100 mice **6.** 750 houses **7.** 72 cars

D

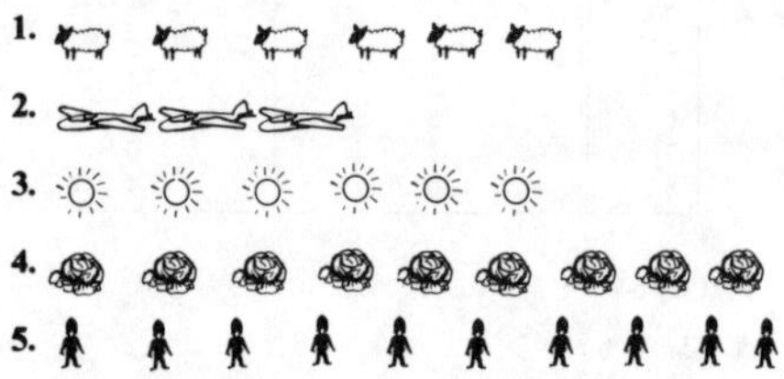

E **1.** 30 cups **2.** 60 kg **3.** 250 litres **4.** 25 chickens **5.** 9 teachers

F Everton 50 Derby 30 Coventry 40 Bolton 35 Arsenal 45
Derby; 5 fewer

PAGE 68

A **1.** 5252·3 **2.** 6724·5 **3.** 9263·7 **4.** 4776·2 **5.** 7452·6 **6.** 8715·1 **7.** 9463·4 **8.** 5327·0 **9.** 0029·9

B **1.** *0·3* **2.** 0·2 **3.** 0·8 **4.** 0·4 **5.** 0·1 **6.** 0·4
7. (1) 0·7 (2) 0·8 (3) 0·2 (4) 0·6 (5) 0·9 (6) 0·6

C **1.** $4\frac{5}{10} \rightarrow 4\frac{1}{2}$ **2.** $5\frac{7}{10}$ **3.** $8\frac{9}{10}$ **4.** $7\frac{2}{10} \rightarrow 7\frac{1}{5}$ **5.** $10\frac{4}{10} \rightarrow 10\frac{2}{5}$ **6.** $10\frac{8}{10} \rightarrow 10\frac{4}{5}$

D **1.** 0·7 **2.** 0·3 **3.** 0·5 **4.** 1·7 **5.** 2·9 **6.** 3·3

PAGE 69

E (practical work)
1. (a) $\frac{25}{100}$ (b) 0·25
2. (a) $\frac{50}{100}$ (b) 0·50
3. (a) $\frac{75}{100}$ (b) 0·75

F
1. 6, 9, 12, 16, 14, 26
2. 4, 7, 9, 12, 5, 13
3. 9, 15, 24, 33, 21, 12

G **1.** *0·5* **2.** *0·05* **3.** 0·2 **4.** 0·02 **5.** 0·3 **6.** 0·03 **7.** $\frac{4}{10}$ = 0·4 **8.** $\frac{4}{100}$ = 0·04 **9.** $\frac{4}{100}$ = 0·04

PAGE 70

A

SMALL SQUARES	FRACTION OF WHOLE	DECIMAL FORM
7	*$\frac{7}{100}$*	*0·07*
28	*$\frac{28}{100}$*	*0·28*
30	$\frac{30}{100}$	0·30
69	$\frac{69}{100}$	0·69
4	$\frac{4}{100}$	0·04
6	$\frac{6}{100}$	0·06
70	$\frac{70}{100}$	0·70
25	$\frac{25}{100}$	0·25
75	$\frac{75}{100}$	0·75
90	$\frac{90}{100}$	0·90

B **1.** *24·35* **2.** 32·46 **3.** 4·02 **4.** 10·50 **5.** 0·08

C

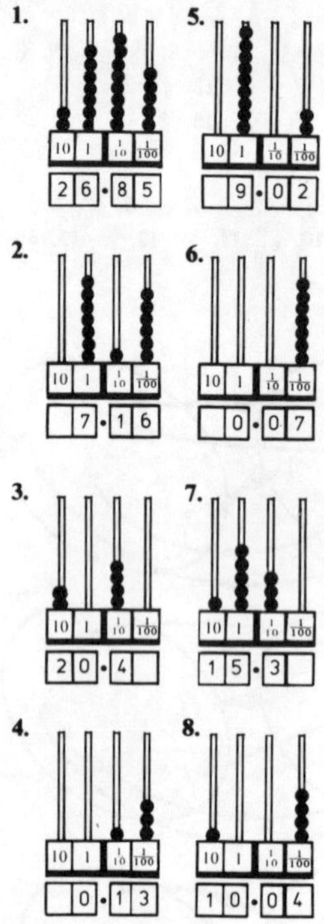

D **1.** Alan **2.** Liza **3.** Eric **4.** Sonia **5.** Jack

PAGE 71

E

DECIMAL FRACTION	IN WORDS	COMMON FRACTION	IN WORDS
7·3	*seven point three*	*$7\frac{3}{10}$*	*seven and three tenths*
5·9	five point nine	$5\frac{9}{10}$	five and nine tenths
3·47	*three point four seven*	*$3\frac{47}{100}$*	*three and forty-seven hundredths*
0·9	nought point nine	$\frac{9}{10}$	nine tenths
2·53	*two point five three*	$2\frac{53}{100}$	two and fifty-three hundredths
5·5	five point five	$5\frac{1}{2}$	*five and a half*
10·09	ten point nought nine	$10\frac{9}{100}$	ten and nine hundredths
1·25	one point two five	*$1\frac{1}{4}$*	one and a quarter
0·93	*nought point nine three*	$\frac{93}{100}$	ninety-three hundredths
6·25	six point two five	$6\frac{1}{4}$	six and a quarter
3·2	three point two	$3\frac{1}{5}$	*three and one fifth*
0·03	nought point nought three	*$\frac{3}{100}$*	three hundredths
0·07	nought point nought seven	$\frac{7}{100}$	seven hundredths
0·75	nought point seven five	*$\frac{3}{4}$*	three quarters

F **1.** *0·4 > 0·2* **2.** 0·3 < 0·8 **3.** 0·5 < 0·9 **4.** 0·06 < 0·6 **5.** 0·7 > 0·07 **6.** 2·0 > 0·7 **7.** 0·15 < 0·51 **8.** 0·6 > 0·09 **9.** 10·2 > 9·5 **10.** 0·07 < 0·1 **11.** 0·4 > 0·15 **12.** 0·9 > 0·85

G **1.** 200, 20, 2, 0·2, 0·02
2. 1500, 150, 15, 1·5, 0·15

PAGE 72

A **1.** 4·13 **2.** 14·98 **3.** 8·66 **4.** 27·47 **5.** 10·10 **6.** 11·00 **7.** 20·01 **8.** 18·03 **9.** 45·00 **10.** 10·94 **11.** 1·01 **12.** 44·71 **13.** 11·26 **14.** 100·00 **15.** 2·35 **16.** 2·06 **17.** 212·2

B
A *48·02* seconds
B 47·95
C 48·31
D 48·71
E 47·43
F 47·97
Team E won.

PAGE 73

C **1.** 22·6 cm **2.** 14·4 cm **3.** 9·64 cm **4.** 14·4 cm **5.** 31·86 cm **6.** 22·0 cm

D
A → D = 12·17 km
B → D = 9·90 km
C → D = 14·95 km
A → E = 11·08 km
B → E = 8·81 km
C → E = 13·86 km
B → E was shortest.

E 2·48 cm

PAGE 74

A **1.** 12·1 **2.** 13·50 **3.** 3·96 **4.** 6·98 **5.** 3·86 **6.** 7·51 **7.** 8·96 **8.** 17·44 **9.** 2·71 **10.** 1·51 **11.** 0·35 **12.** 0·17 **13.** 0·58 **14.** 6·91 **15.** 4·91 **16.** 1·97 **17.** 6·6

B
1. 24·7 15·63
2. 15·31 7·07
3. 0·5 0·41
4. 0·7 0·49
5. 1·03 0·94
6. 2 1·3
7. 15·2 5.2
8. 10 8·48

C **1.** 97·55 m
2. 25·37 m shorter
3. 169·73 m altogether

PAGE 75

D **1.** 494·7 **2.** 509·2 **3.** 353·5 **4.** 219·7 **5.** 522·6
No. 4 had used least, no. 5 most.

E **1.** (a) 119·2 km
(b) 415·9 km
(c) 572·7 km
(d) 145·8 km
(e) 278·7 km
2. 1532·3 km

F **1.** (a) 781 km
(b) 19·4 km
2. £3·95
3. 0·63 m

PAGE 76

A **1.** (a) (A,4)
(b) (B,3)
(c) (C,4)
(d) (A,2)
2. (a) Berg
(b) Ming
(c) Ham
3. *(A,4)*, (A,3), (B,2), (C,2), (C,1)

B **1.** Edna (*6*,*3*) Bob (4,2) Stan (5,2) Lesley (1,3) Peggy (2,1) Irene (3,3) Harry (6,1) Margery (3,4)
2. (a) Tom
(b) Brenda
(c) Amanda
(d) Susan
(e) Tony
(f) Lynn
(g) Nick
(h) Jean
3. (5,3)

PAGE 77

C A (2,0) B (8,1) C (10,0) D (9,1) E (10,3) G (4,3) I (2,2) J (0,2)

D

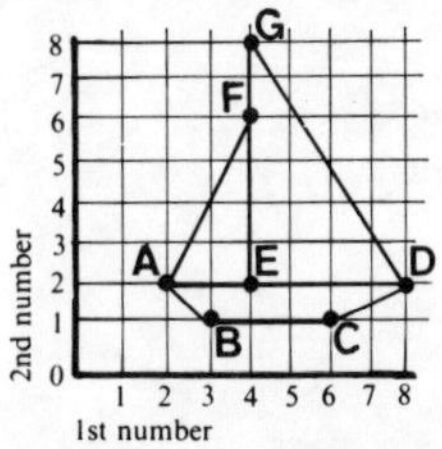

E

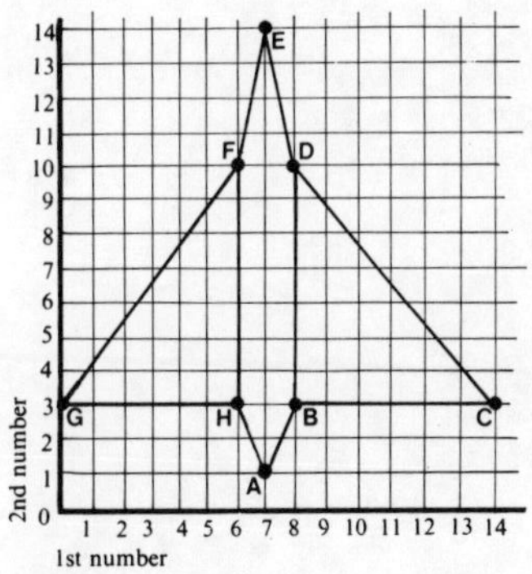

F

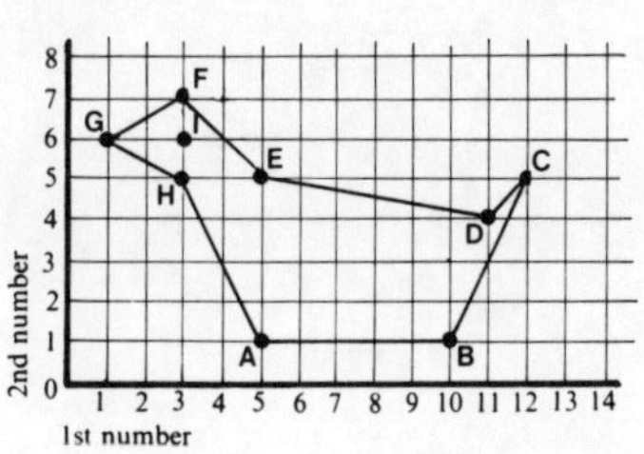

PAGE 78

A **1.** 1 **2.** 10
1 14
1 18
1 22

B **1.** 348 **2.** 127 **3.** 812 **4.** 78 **5.** 302 **6.** 681 **7.** 75 **8.** 973 **9.** 768 **10.** 1842 **11.** 686 **12.** 13 **13.** 13 **14.** 81 rem 2 or $81\frac{2}{3}$ **15.** 91 rem 4 or $91\frac{1}{2}$ **16.** 50 rem 7 or $50\frac{7}{9}$

C **1.** *100 cm is equal to 1* metre *and 1 cm = 0·01* metre.
2. 10 *mm is equal to 1 cm and 1 mm = 0·1 cm.*
3. 100*p is equal to £1 and 1p =* £0·01.
4. *60 seconds is equal to 1* minute *and 30 seconds* = 0·5 minute.
5. *7 days is equal to 1 week and 1 day = $\frac{1}{7}$ week.*

D **1.** 566 g
2. (a) 1000 g flour, 100 g lard, 8 g sugar, 8 g salt, 16 g yeast $\frac{1}{2}$ litre water
(b) 250 g flour, 25 g lard, 2 g sugar, 2 g salt, 4 g yeast, $\frac{1}{8}$ litre water

PAGE 79

E **1.** 28 cm^2 **2.** 30 cm^2 **3.** 36 cm^2
No. 3 is a square.

F **1.** (a) $\frac{1}{5}$ (b) $\frac{1}{5}$ (c) $\frac{1}{10}$ (d) $\frac{7}{20}$ (e) $\frac{1}{2}$
2. (a) $\frac{20}{100}$ (b) $\frac{20}{100}$ (c) $\frac{10}{100}$ (d) $\frac{35}{100}$ (e) $\frac{50}{100}$
3. (a) 0·20 (b) 0·20 (c) 0·10 (d) 0·35 (e) 0·50

G **1.** $\frac{3}{4}$ **2.** $\frac{1}{2}$ **3.** $\frac{7}{10}$ **4.** $\frac{5}{8}$ **5.** $\frac{7}{12}$ **6.** $\frac{3}{8}$ **7.** $\frac{1}{2}$ **8.** $\frac{1}{3}$ **9.** $\frac{1}{10}$ **10.** $\frac{3}{8}$

H **1.** M (2,3), N (4,3), T (3,2)
2, 3.

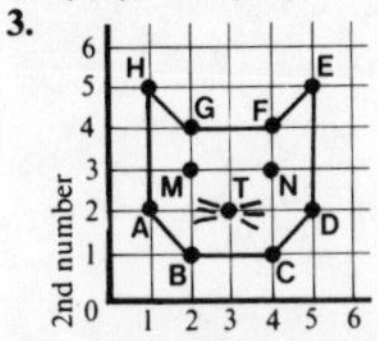

I **1.** 24·58 **2.** 8·66 **3.** 18·87 **4.** 38·59 **5.** 47·98 **6.** 9·38

PAGE 80

A (individual work)

B

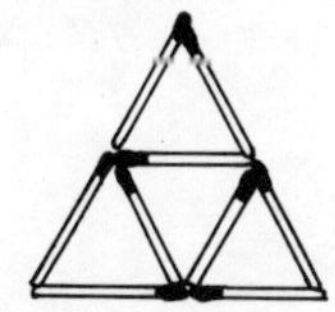

18 matches → 13 triangles

C $\frac{1+3}{3+3} \rightarrow \frac{4}{6} \rightarrow \frac{2}{3}$

D

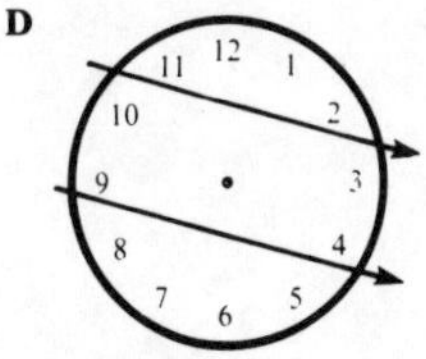

E **1.** 5 cm, 1 cm **2.** 12 cm **3.** 4 more rectangles **4.** 25 cm^2